[redacted]

FUNDAMENTAL ASTROMETRY

FUNDAMENTAL ASTROMETRY

DETERMINATION OF STELLAR COORDINATES

V. V. PODOBED

English Edition Edited by A. N. VYSSOTSKY

THE UNIVERSITY OF CHICAGO PRESS
CHICAGO AND LONDON

Russian edition published by Fizmatgiz, Moscow, 1962.

Translated from the Russian by Scripta Technica, Inc.

Library of Congress Catalog Card Number: 64-15810

THE UNIVERSITY OF CHICAGO PRESS, CHICAGO & LONDON
The University of Toronto Press, Toronto 5, Canada

Published 1965. Printed in the United States of America.

Editor's Preface

The author of this book is Docent in the school of Mathematics of the University of Moscow where he received his training in Astronomy. As stated in the closing chapter, he has attempted to present an elementary outline of various problems related to the determination of accurate stellar positions. The translation follows the original text closely.

I wish to thank Dr. Chester B. Watts and Mr. Francis P. Scott, astronomers of the U.S. Naval Observatory, who were kind enough to supply certain technical terms, and the author himself, who furnished the transliteration of several unfamiliar proper names.

At the end of the Russian text is a list of books and articles for further consultation. However, since nearly all of these are in Russian, I have compiled a similar list of references in English or German.

A. N. V.

Preface

In order to set up a basic coordinate system in the celestial sphere, fundamental astrometry solves two problems: it determines the coordinates of the stars and establishes the fundamental astronomical constants. In the present text, the first of these problems, that of determining the coordinates of the stars, is expounded.

In astronomical literature there is no systematic elementary description of the principles of the theory and their application in the determination of the absolute coordinates of the stars. A significant survey, *Fundamental'naya astrometriya* [Fundamental Astrometry], by M. S. Zverev (corresponding member of the Academy of Sciences of the USSR), published in *Advances in Astronomical Sciences* (vols. V and VI), is designed for trained specialists and, because of its complicated nature, cannot be used as a textbook.

The present course is designed for student astronomers specializing in astrometry and for astronomers who are beginning their studies in the field of meridian astrometry. It is based on a course of lectures given at Moscow State University.

The author hopes that this course will also be useful to professional astronomers of all specialties, to geodesists, and to amateur astronomers.

The problems of determining the coordinates of stars by classical methods with meridian instruments constitute the main part of the course. Since photographic methods are also used for solving a number of fundamental problems of astrometry the essentials of photographic astrometry are presented. In the last part, the course contains a description of the methods of deriving the fundamental coordinate system and a survey of the principal catalogs of stellar positions and motions. The fundamental astronomical constants and their determination are not included in the course because of the special characteristics of this problem. A monograph, *Fundamental'nye postoyannye astronomii* [Fundamental Astronomical Constants], by Professor K. A. Kulikov is devoted to this question and fills this gap very well.

Since the author studied at the Moscow School of Astrometry, in presenting the course he has followed the teaching methods of Professor

S. N. Blazhko and M. S. Zverev at Moscow State University. He has tried to present only the essentials of the problems and not to provide solutions for every actual case, leaving the reader to draw his own conclusions. The author sees in this a stimulus to further study of the material discussed. At the same time, he has tried not to burden the course with superfluous details, keeping in mind the main problem, namely, that of imparting the fundamentals of the subject to the beginning astrometrist. For this reason, mathematical computations have been kept within reasonable limits.

At the end of the book there is a list of recommended literature.*

The author expresses his deep gratitude to Corresponding Member of the Academy of Sciences M. S. Zverev, to Candidate of the Physical and Mathematical Sciences A. P. Gulyaev, and to Junior Scientific Worker T. S. Meshkova, all of whom familiarized themselves in detail with the manuscript and whose advice helped in many ways. The author also expresses gratitude to Professor D. Ya. Martynov and Senior Scientific Worker P. I. Bakulin for their comments and to B. K. Bagil'dinskiĭ and N. A. Zaĭtseva for providing the photographs and drawings for the book.

Moscow State University,
Sternberg State Astronomic Institute,
Department of Astrometry

V. V. Podobed

*A modified list has been compiled for the use of readers of the English translation of this book. Ed.

Contents

Contents

CHAPTER I

Fundamental Astrometry and Its Basic Problem—Determination of the Coordinates of Stars

1. The Subject of Fundamental Astrometry and the Concept of an Inertial Coordinate System

It is possible to study the radiant energy which reaches us from distant heavenly bodies in two ways, either by analyzing its physical properties, or by determining the precise direction from which it approaches the earth. Astrophysics solves problems associated with the characteristics of the light from heavenly bodies, whereas astrometry is concerned with questions of the directions of the bodies, the change in these directions and angular distances of the bodies from one another. Astrometry is the science dealing with the theory and application of geometric measurements in astronomy and with the establishment of celestial coordinates. The basic mathematical apparatus for solving problems of astrometry is provided by spherical astronomy, the theoretical bases of which are closely tied up with celestial mechanics.

Astrometry is divided into fundamental astrometry and practical astronomy, the latter including the determination of geographic coordinates, of time, and of latitude. Fundamental astrometry is the science whose chief problem is the construction on the celestial sphere of an absolute coordinate system. Fundamental astrometry is also concerned with the determination of the basic astronomical constants, that is, the quantities that allow us to compute the regular changes of the coordinates with time.

A fundamental system of coordinates is fixed in the celestial sphere in the form of a general catalog of stars. For a certain number of stars distributed over the celestial sphere, this catalog contains the right ascensions and declinations and also the proper motions and precessional quantities that allow us to transfer the mean coordinates from one equinox to another. To know the apparent positions of the stars for an arbitrary epoch, it is necessary to know the fundamental astronomical constants.

The construction of a system of celestial coordinates and the determination of the fundamental constants are impossible without a knowledge of the coordinates of the stars. Since approximate coordinates are known for the majority of stars which interest astrometrists, the problem becomes one of making small corrections to these coordinates. The division of fundamental astrometry dealing with the theory and application of the determination of the coordinates of the stars is often called *meridian astrometry* because observations of the stars for determining the coordinates by classical methods are carried out on the meridian.

The problem of meridian astrometry is not limited to the determination of coordinates of the relatively small number of stars necessary for the construction of a fundamental coordinate system. Practical, photographic, and stellar astronomy are inconceivable without a knowledge of the coordinates and proper motions of a large number of both bright and faint stars. The catalogs necessary for all these problems are constructed from meridian observations. We may therefore assert that meridian astrometry gives a vital basis to the solution of all astrometric problems.

The absolute system of coordinates used as a basis for mechanics has been considered since Newton's time as being bound to a system of "motionless" stars and, consequently, as being motionless in space. But only in theory can a simple coordinate system be considered motionless. The development of astronomy has shown that motionless stars do not exist; furthermore, besides the individual motions of the stars, there are regular aggregate motions. Therefore, astrometrists trying to set up a system of coordinates have been confronted with the problem of constructing a system of celestial coordinates that would allow them to solve problems associated with movements in the universe and particularly in our galaxy.

A system of celestial coordinates suitable for this purpose should ideally be inertial; that is, it should possess only rectilinear, uniform motion with no rotation. To get such an ideal inertial system, one must consider the motion of the observer in space and the motion of those celestial objects to which the system of celestial coordinates is bound.

Observations of celestial objects (stars) take place upon the earth, which has certain finite dimensions and, most important, has a complex motion in space. All this makes impossible direct comparison and coordination of observations conducted on earth at different points and at different epochs. The problem reduces to one of transfer from apparent topocentric coordinates obtained from observations to mean heliocentric coordinates. This can be done by considering the effects of such phenomena as precession, nutation, aberration, and parallax. As a result of these reductions, the catalog contains the positions of the stars for a certain epoch in a system of coordinates with its origin at the center of the sun and moving with it. This holds for every catalog obtained from observations with a given instrument.

If the stars and the sun did not possess their own individual proper motions and did not have aggregate motions, the coordinate system obtained from observations could be used for any epoch because the motion of the earth, despite its complex nature, can be taken into account. But the sun and the stars do move; we must therefore determine the proper motions of the stars. Heliocentric coordinates and the proper motions of a certain number of stars afford, in practice, a system of celestial coordinates which is inertial to an accuracy limited by our knowledge of the motion of the sun and the aggregate motions of these stars.

The study of the proper motions and of the radial velocities of the stars has demonstrated the existence of solar motion in space and the rotation of the stellar system that is our galaxy. We should note that the present-day accuracy of the majority of practical astrometric problems does not demand taking solar motion or the rotation of the galaxy into account. Until quite recently, the calculation of the influence of these factors on the system of celestial coordinates has been within the competency of stellar astronomy, which deals with refined ideas concerning the motion of the stars.

Secular aberration, which is the result of the motion of the sun, is not generally taken into account because its influence can be considered as constant with time and, consequently, constant for every star.

The idea, advanced in the last few decades, of tying the system of celestial coordinates by purely astrometric methods to distant nebulae outside the galaxy, that is, to practically motionless objects, will enable us to obtain a more perfect inertial system of coordinates independent of the motion of the sun or rotation of the galaxy. This will also make possible a detailed study of the motions in the galaxy from an essentially new point of view. The construction of such an inertial system is a matter for the coming decades.

2. The General Principles for Absolute and Relative Determinations of the Coordinates of Stars from Meridian Observations

As a basic system of celestial coordinates, we take an equatorial system, which is related to the possibility of exact fixation in space of certain planes whose positions are given by natural phenomena. Since we make our observations from an earth which rotates and moves in space, the planes defined by the daily rotation and the annual revolution of the earth around the sun—the equator and the ecliptic—are the most convenient planes for this purpose. As we know, a spherical system of coordinates is uniquely defined by a great circle and a point on it. The equator and one of the points of its intersection with the ecliptic—the point of the vernal equinox—can similarly be used for an equatorial system. A change in the positions of the equator and the ecliptic in space takes place as a result of the gravitational interaction between the earth, the moon, the sun, and the planets. Therefore, calculation of the positions of these planes is possible for any epoch if the astronomical constants are empirically determined from observations. The coordinate system itself is fixed on the celestial sphere by the spherical coordinates obtained from observations and by the proper motions of a certain number of stars which are as uniformly distributed throughout the sky as possible.

The equatorial coordinates—*the right ascension and declination*—of the stars are determined by the classical method from observations of the culminations of the stars. Meridian instruments are oriented with respect to a plumb line, a fact that makes observation and correction for instrumental errors easier. The suitability of meridian observations derives from the fact that a star crosses the meridian parallel to the horizon, without changing its altitude.

On the meridian, the circle of declinations and the circle of altitudes coincide. Therefore, right ascension α and declination δ are individually connected by simple relations with the observed instant of passage across the meridian T and with the zenithal distance on the meridian z. This makes it possible to determine independently the right ascension from the observed values of T and the declination from the observed values of z. The right ascension of the star on the meridian is equal to the local sidereal time S at the upper culmination or differs from it by twelve hours at the lower culmination. The declination of the star on the meridian is connected with the zenithal distance and the local latitude, φ, by the relation $\delta = \varphi \pm z$ for the upper culmination, and $\delta = 180° - \varphi - z$ for the lower culmination (the plus sign is used for a culmination to the north of zenith and the minus sign for a culmination to the south of zenith).

The observations of T and z can be made separately or together. In the second case, both coordinates α and δ are obtained from observations on the same evening.

There are two methods of determining the coordinates: the *absolute* and the *relative*. The absolute method proposes determination of the coordinates of the stars without using the stellar coordinate points from previous determinations. In the relative method, the positions of the stars are obtained relative to stars with known, accepted coordinates.

The problem of determining the declinations by the absolute method becomes one of measuring the zenithal distances of the stars. In fact, it follows from the formulas given above that, to compute the declination, we must know the zenithal distance of the star and the latitude. The latter is easily determined from observations of circumpolar stars in two culminations. If we take the difference in the expressions for the two culminations of a single star, its unknown declination can be eliminated, and, for determining the latitude φ, we obtain the relation

$$\varphi = 90° - \frac{1}{2}(z_N + z_V),$$

where z_N and z_V are the zenithal distances of the circumpolar star at the upper and lower culminations respectively. Thus, the latitude can also be computed in terms of observed zenithal distances. The zenithal distances can be determined by means of an instrument having a graduated circle mounted parallel to the plane of the meridian.

When determining right ascensions with an astronomical clock, the instants at which a star crosses the meridian are noted. If we denote the correction of the clock by u, we can write for the observation of each star

$$\alpha_i = S_i = T_i + u_i.$$

If we take the difference for two stars, we get

$$\alpha_i - \alpha_j = (T_i - T_j) + (u_i - u_j),$$

where T_i and T_j are obtained from observations and

$$u_i - u_j = \frac{\omega}{24}(T_i^h - T_j^h).$$

The diurnal rate of the clock ω is determined from the difference between the instants of passage across the meridian of the same stars on successive nights, which allows us to eliminate their unknown right ascensions.

Observation of the passage across the meridian of only a few stars will give us the differences in their right ascensions because the zero point of right ascension, or the point of the vernal equinox (which is determined by the intersection of the planes of the ecliptic and the equator, and is dependent on the motion of the earth) is in no way marked on the celestial sphere; therefore, it is impossible to observe the instant of its culmination. To determine the position of the point of vernal equinox in the sky relative to the stars, it is necessary to observe the sun, the motion of which determines the plane of the ecliptic. From the observed declination of the sun $\delta_\odot$, it is possible to compute its right ascension $\alpha_\odot$. If we know from observations the difference between the right ascensions of the sun and of some star $\alpha_i - \alpha_\odot$, we can obtain the right ascension of the star. This is only the principle; in practice the matter is much more complicated.

Relative determinations of the coordinates amount to a determination of the differences of the coordinates of the stars whose positions are to be determined and of reference stars; the coordinates of the latter are assumed to be known. The difference in the readings of the graduated circle in the plane of the meridian at the time of observation of the reference star and of any other star is, after correction for refraction, the difference in the declinations of these stars:

$$M_{\text{ref}} - M_{\text{det}} = \delta_{\text{ref}} - \delta_{\text{det}} = \Delta\delta.$$

The difference of the moments of transit across the meridian, after taking into account the "run" of the clock, is the difference of the right ascensions:

$$T_{\text{ref}} - T_{\text{det}} = \alpha_{\text{ref}} - \alpha_{\text{det}} = \Delta\alpha.$$

From these relations, knowing α_{ref} and δ_{ref} and the observed values of $\Delta\alpha$ and $\Delta\delta$, we can easily obtain the desired coordinates α_{det} and δ_{det} for a particular star.

All the reasoning in this section is valid for an ideal instrument without error and ideally oriented in space. Under actual conditions, these simple geometric principles are greatly complicated by the presence of random and systematic errors of observation, the elimination of which causes astrometric observation to take on the nature of a craft and even of an art.

3. Random and Systematic Errors in Observations

The results of meridian observations made with various instruments at different observatories are compiled in the form of stellar catalogs, absolute or relative, depending on the method of observation used. These catalogs contain, for the epoch of observation, the right ascensions and declinations of the stars of an appropriately chosen list, and serve as a source of material for solving all problems of fundamental astrometry and especially for establishing a fundamental system of coordinates. This last problem amounts to deriving an equatorial system of coordinates for some epoch as well as the proper motions of the selected stars. Absolute catalogs are used for setting up a coordinate system and a system of proper motions; relative catalogs are used only for reducing random errors in coordinates and in proper motions of the stars.

A fundamental system of coordinates that, in a certain sense, is the most probable one, should be as free as possible from the random and systematic errors of the original catalogs and should not possess a rotational motion; that is, it should be inertial. The question of reducing observed coordinates to an inertial system was explained in Section 1. Here we examine the theoretical question of the influence of random and systematic errors of observation, the last of which has caused differences in the coordinate systems of the initial catalogs and has been the source of the greatest inconveniences to astrometrists.

Every catalog obtained from observations by means of various instruments has both random and systematic errors. Random errors are traceable to accidental causes and are estimated on the basis of the probability theory. They can be reduced by increasing the number of observations.

Systematic, as opposed to random, errors persist in successive observations under analogous conditions. At present, systematic errors in the catalogs constitute the basic obstacle to increasing the accuracy of celestial coordinates. Therefore it is natural that astrometrists' efforts should be primarily directed to overcoming these errors.

There are three kinds of random and systematic errors in observation: instrumental errors, errors caused by changes in external conditions, and personal errors of the observer.

(a) Instrumental errors are a consequence of defects in the meridian instrument. They are subdivided into errors made in setting up the instrument (collimation, inclination, and azimuth of the horizontal axis, variation in the clock's rate, etc.) and imperfections in the instrument itself (flexure, imperfections in the pivots, errors in the graduated circle, etc.). It is impossible to make a geometrically ideal instrument or to set it up exactly. Therefore, an investigation of the instrument for determining its errors is absolutely imperative to the performance of any astrometric work. A suitably chosen observation procedure can reduce the influence of instrumental errors.

Although instrumental errors are relatively small, the uncertainty in determining the individual errors, which is often dependent on ignorance of their causes, has a considerable influence on the accuracy of contemporary catalogs. A striking example is flexure of the telescope tube. Calculation of this error, despite the numerous studies made of it, has not yet been completely successful, with the result that there are basic discrepancies in the catalogs of the declinations of the stars.

The theory of an astrometric instrument is basically the theory of its errors and of the methods of reducing their influence. Astrometrists devote no less study to instrumental errors than to the observations themselves.

(b) Variation in external conditions, both from season to season and from hour to hour, is another reason for the existence of observational errors. The chief role is played by variation in temperature, which causes deformations of the various parts of an instrument and, by acting on the foundation and the supports of the instrument, influences its stability. Variation in the characteristics of the air surrounding the instrument—its temperature and pressure, velocity and direction of the wind—can cause an anomalous refraction and lower the quality and clarity of the images, which increases both accidental and systematic errors of observation. Calculation of the influence of external conditions is rather difficult as a result of the lack of knowledge of how they affect the quantities observed.

In practice, the calculation of the influence of external conditions is limited to studies of the behavior of instrumental errors in the course of an evening, or from season to season, as well as to some attempts to correlate the deviations obtained in observations with the characteristics of the air in the pavilion, near the surface of the earth, in the upper layers of the atmosphere, and with the temperature gradient in the instrument.

(c) Personal errors are the result of the psycho-physical characteristics of the observer. They are reduced by the application of new methods of observation, in particular of photoelectric methods, which eliminate the eye of the observer from the measurements. Increasing the number of observers helps decrease the effect of personal errors.

Diminishing the influence of accidental and, especially, of systematic errors on observations, leading to an improvement in the accuracy of stellar catalogs and, consequently, to an improvement in the fundamental system, determines the lines of development of the technology and methodology of astrometric observations. These considerations are involved in the perfection of contemporary instruments, the development of new principles of determination of coordinates, the construction of new instruments for realizing these principles, and the application of new methods of bringing stellar catalogs into agreement.

4. A Short History of the Development of the Instruments Used in Fundamental Astrometry

From the principles of determining coordinates that we have been examining, it follows that, to carry out observations, one must have:

(1) An instrument that makes it possible to observe celestial bodies as they cross the celestial meridian and to measure the angles in the meridian;

(2) An astronomical clock for registering the instants of crossing.

A graduated circle is a scale for measuring declinations and an astronomical clock is a scale for measuring right ascensions.

The determination of coordinates is made by the classical method with three instruments: a transit instrument (which enables us to obtain the right ascensions), a vertical circle (which measures the declinations), and a meridian circle (suitable for determining both coordinates). Contemporary methods for determining coordinates, like contemporary astrometric instruments, have not been developed overnight but have come about as the result of a long process of development.

In the early stages of astrometric development, determination of the coordinates of celestial bodies was the primary problem of ancient astronomers, who clearly grasped the practical value of studying the motions of celestial bodies and predicting their positions on the celestial sphere.

Of the earliest instruments used in ancient times and in the Middle Ages for determining the coordinates of celestial bodies, we should mention the armillary sphere and the quadrant. These instruments had wide distribution: observations were made with them in China, India, Europe, and the Near East.

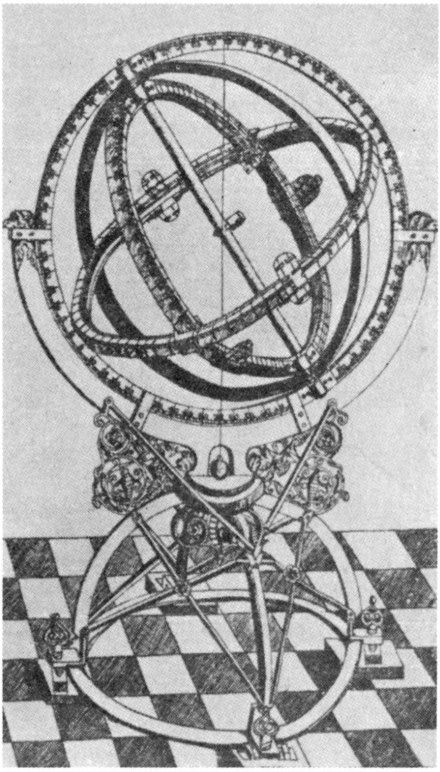

Fig. 1. The armillary sphere of Tycho Brahe (16th century).

The armillary sphere (Fig. 1) was a model, as viewed from the outside, of the celestial coordinates, and consisted of concentric circles that could be placed in a position parallel to the basic circles of the celestial sphere. For sighting stars or planets, astronomers used diopters consisting of two disks placed at opposite ends of the diameter, one with a slit and the other with cross threads. The positions of the diopters were read from an index. The ecliptical coordinates (the latitude β and the longitude λ) were determined by means of armillary spheres. The first stellar catalogs, obtained by means of armillary spheres, had an accuracy of the order of $\pm 15'$.

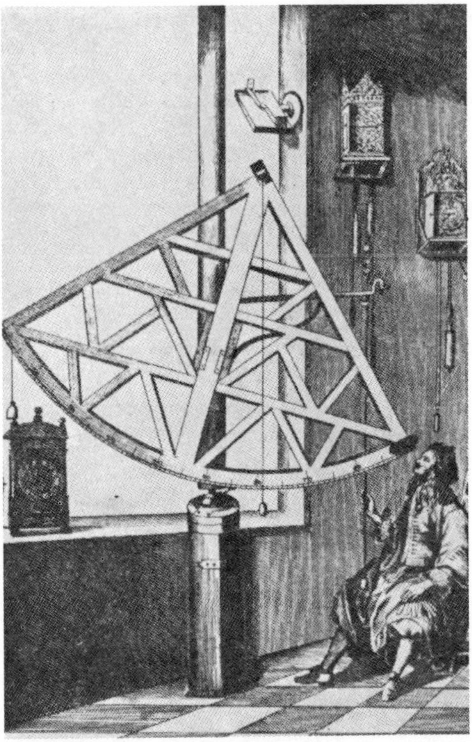

Fig. 2. The quadrant of Johannes Hevelius (17th century).

The difficulty of precisely orienting the circles of an armillary sphere caused astronomers eventually to prefer quadrants (Fig. 2). A quadrant was a quarter of a circle and was placed in a vertical plane. It was either stationary in a meridian (often on a stone wall—a wall quadrant) or it could be rotated on an azimuthal mounting. In the case of the wall quadrant, two diopters were placed on an alidade, that is, a straightedge extending along the radius of the quadrant and having its axis of rotation at the center of the arc of the quadrant. The quadrant was set up vertically, and with it one could measure the altitudes of heavenly bodies and note the instants of their transit of the meridian. We must not omit to mention the enormous sextant, similar in principle to the quadrant, that was devised by the Uzbek astronomer Ulugbek for observing the sun; it was located in a trench dug in the earth and had a radius of about forty meters.

The science developed in the Middle Ages for instrument-making increased the accuracy obtainable for coordinates to $\pm 2'$.

In the seventeenth century, a whole series of improvements in astronomical instruments was made. In the sixteen thirties, P. Vernier devised a means of reading the circle more accurately (the vernier). In the forties, W. Gascoigne invented the ocular micrometer, and in the sixties, J. Picard first used a tube as a sighting device. Pendulum clocks were beginning to be used in observations.

The development of celestial mechanics on the basis of the discovery of Kepler's laws demanded greater accuracy in determining coordinates. The quadrant was now unable to satisfy the growing demand because of many structural imperfections.

O. Roemer is considered the "father" of the modern classical meridian instrument. In 1689, he constructed the prototype of a transit instrument and later the meridian circle (Fig. 3).

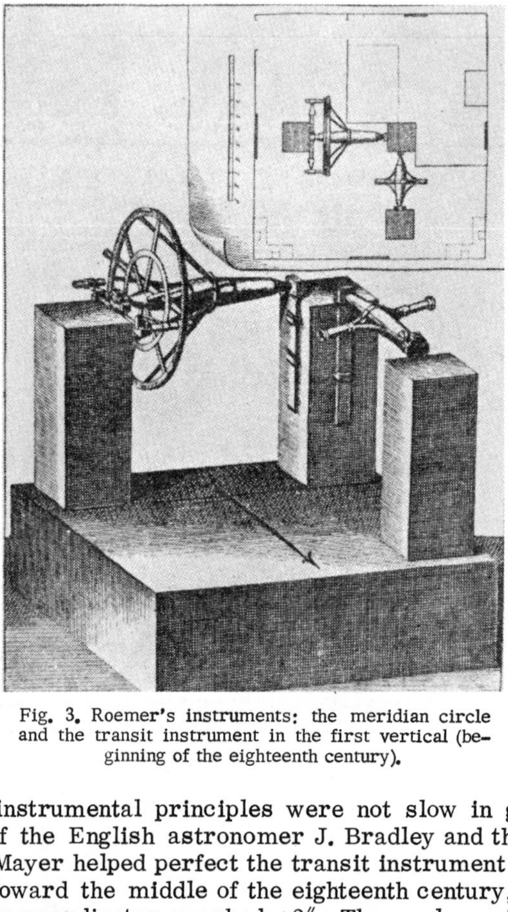

Fig. 3. Roemer's instruments: the meridian circle and the transit instrument in the first vertical (beginning of the eighteenth century).

The new instrumental principles were not slow in giving results. The efforts of the English astronomer J. Bradley and the German astronomer T. Mayer helped perfect the transit instrument and developed its theory. Toward the middle of the eighteenth century, the accuracy in determining coordinates reached $\pm 2''$. The ocular micrometer was perfected. At the beginning of the nineteenth century, the complete graduated circle replaced the quadrant altogether. This eliminated eccentricity errors. Microscope-micrometers came into use for reading the circles. The meridian circle of the modern type was created. Along with the development of the technical aspects, we should also mention the development of methods of observation and processing data.

The beginning of contemporary fundamental astrometry might be considered as coinciding with the beginning of the work of the Greenwich Observatory, founded in 1675. Although the first catalogs of the Greenwich Observatory do not at present have any scientific value because of their poor accuracy, the general organization of the work of the Greenwich Observatory made possible the development of fundamental astrometry in subsequent years. Regular astronomical observations have been carried on at Greenwich for more than two hundred years. Numerous stellar catalogs have been compiled as a result of these observations. Of special value are the series of observations of the moon made at Greenwich.

Fig. 4. Friedrich Wilhelm Bessel (1784-1846).

We should note that after the flowering of the Greenwich school of astrometry in the eighteenth century, brought about by the works of J. Flamsteed, N. Maskelyne and J. Bradley, the accuracy of the Greenwich catalogs remained almost constant until the middle of the nineteenth century. It became clear that, to increase the accuracy of the coordinates obtained, new principles of observing and processing were necessary. These new principles were worked out by F. W. Bessel and V. Ya. Struve in the first half of the nineteenth century.

By his many-sided investigations, Bessel virtually brought the theory of errors of the meridian instrument almost to its present-day status. The path that Bessel pointed out for increasing the accuracy of the catalogs was brilliantly followed in the work of the Pulkovo Observatory.

Immediately after the Pulkovo Observatory was put in service (1839), its founder, V. Ya. Struve, set up new principles as the basis of work in the field of fundamental astrometry. The principle that Struve adopted

Fig. 5. Vasiliy Yakovlevich Struve (1793–1864).

for separate determination of coordinates with perfected specialized instruments (a transit instrument for right ascensions, and an instrument he devised for declinations—a vertical circle) was strikingly justified. The development of the theoretical bases of fundamental astrometry, painstaking investigation of the instruments and computation of their errors, a rational organization of the observations and data processing—these are what immediately led the Pulkovo Observatory to occupy an outstanding place among the astronomical observatories of the world. The importance of the Pulkovo catalogs exceeded that of the catalogs of such long-famous observatories as those of Greenwich and Paris, to say nothing of the remaining observatories. The quality of the Pulkovo catalogs brought fame to the Pulkovo Observatory, which was referred to as the astronomical capital of the world.

Astronomical observational techniques and methods were perfected in the course of the development of astrometry in the second half of the nineteenth centrury. The introduction of the recording chronograph in the sixties and of the registering micrometer of Repsold at the end of the nineteenth century, and the technical improvement of the instruments and auxiliary apparatus brought the accuracy of determining coordinates, with respect to random errors, to the present level ($\pm 0''.40$). All the efforts of astrometrists began to be directed towards overcoming the systematic instrumental errors (errors in the graduated circle, imperfections in the pivots, flexure of the telescope tube, etc.).

Fig. 6. Main building of the Pulkovo Observatory.

Further perfections in the meridian instruments, especially in the last few decades, have been associated with the application of the latest technical achievements: mechanical guidance of the threads of the ocular micrometer and photography of the graduated screw heads, photoelectric registration of transits, and photographic readings of the graduated circle. We should note that the meridian instruments used at the present time are made of steel in contrast with the brass of former times. The construction of pendulum clocks was perfected on the basis of the principle of the free pendulum, which greatly increased their accuracy. Electronics is taking a greater and greater place in astronomical instruments. Quartz clocks, quartz printing chronographs, and electronic computers are a few examples.

At the present time, we can hardly expect more than a two- or threefold increase in the accuracy of determining coordinates, with respect to random errors, by means of instruments based on Roemer's principles. It may be possible to decrease the systematic errors somewhat by perfecting the techniques of construction and control of an instrument, but they can hardly be decreased by an order of magnitude.

In connection with this, investigators have naturally sought new methods and principles of observation and, consequently, have derived essentially new instruments. The most noteworthy change in the principles of determining coordinates has been the shift from observations in the meridian to observations at equal altitudes. Among the instruments that work on this principle should be mentioned the impersonal prismatic astrolabe of A. Danjon (French) and the circumzenithal, developed by E. Buchar in Czechoslovakia. However, these efforts have not given altogether definite results and therefore it is not clear whether this is an advantageous trend. Attempts to construct horizontal meridian instruments are of great interest.

At the present time, classical meridian instruments, at the height of their possibilities, still remain the basic, widely used instruments of fundamental astrometry.

After this short introduction, the purpose of which was to acquaint the reader with the range of problems examined in fundamental astrometry, we go on to a systematic exposition of the course.

CHAPTER II

General Description of Classical Instruments Used in Fundamental Astrometry

5. The Telescope as a Sighting Device

In addition to increasing the amount of light falling on the eye, another important property of the telescope is used in astrometric instruments, namely, that of fixing the direction in space or of serving as a sighting device. In fundamental astrometry, the telescope is used for sighting the objects to be observed and also for the control and investigation of the instrument. In the last case, an auxiliary telescope called a collimator gives the direction along which the instrument is to be aligned.

The telescope was first used as a sighting device in the seventeenth century. This sharply increased the accuracy of sighting. The advantage of the telescope over the diopters previously used consists in the increase in the angular scale, which in turn increases the resolving power of the eye. Therefore, the alignment of the cross threads with the image of the object sighted can be done with much greater certainty than can the combination of the slit and thread of diopters with the object seen by the naked eye.

The telescope used in astrometry consists of an objective, which produces an image in the focal plane of an ocular micrometer with a grid of threads for making measurements in the focal plane, and an eyepiece which serves to increase the accuracy of sighting.

Ordinarily, the objective of a telescope is achromatic and consists of two lenses. The simplicity of its construction depends upon the small field of vision used for visual observations, which rarely exceeds one degree. Thus, we need not make great demands for optical quality and, in practice, the errors of the objective in visual astrometry play a secondary role. It is important only that the quality of the image be sufficiently good near the optical axis.

The objective is characterized by two linear quantities, namely, the diameter of the aperture, and the focal length. For simple objectives, the diameter of the aperture may be considered as coinciding with the diameter of the mounting of the front lens. The focal length F is determined by the distance from the secondary principal point of the objective to the main focus; that is, the point at which light rays parallel to the optical axis converge after falling on the objective. The plane passing through the main focus perpendicular to the optical axis is called the main focal plane.

A stationary grid of threads tightly connected with the objective by a metallic tube is placed in the focal plane of the telescope objective. The line connecting the secondary principal point of the objective with the central cross of the network of threads is called the line of sight. This line fixes the direction in space parallel to the direction of the straight line connecting the primary principal point of the objective and the point of the object whose image is on the central cross of the threads in the field of vision. To increase the accuracy of alignment of the image and the crossed threads, the observer looks through an eyepiece. The greatest magnification most commonly used is about two or three hundred times.

Parallel light rays falling on the objective at a small angle ι with the optical axis converge into a focal point lying in the main focal plane. The distance of this point from the optical axis is $l = F \tan \iota$. Since these angles ι are small in visual astrometry, we can consider l proportional to ι: $l = F\iota$. This relation between the line segment, l, in the focal plane and its angular dimension, ι, is used to measure the angles in the field of vision of the telescope.

To measure the linear distances in the focal plane and consequently the corresponding angular distances on the celestial sphere, a filar micrometer is used, where the linear displacement of the thread is proportional to the angle of rotation of the micrometer screw. Since contemporary instruments are brought to bear on the observed objects by the moving thread of the micrometer, there are ways of transferring from the direction fixed by the micrometer thread to the direction set by the sighting line.

The angular value ι, corresponding to the displacement of the movable thread of the micrometer when the graduated head of the micrometer is rotated n times, is easy to calculate if we introduce the concept of a single turn of the screw of the micrometer R. The value of the turn is the arc on the celestial sphere corresponding to the displacement of the movable thread for a turn of the screw. Then, $\iota'' = R \cdot n$, where n is expressed in turns or parts of a turn of the screw, and R is the amount in seconds of arc per turn. All astronomical measurements are based on this relationship. The value of R is determined from special investigations.

6. The Transit Instrument and Its Basic Formula

As was shown in Section 2, in order to determine the difference between the right ascensions of stars, we should be able to register the instants of their transits in the plane of the celestial meridian.

For observing the passages of the stars, we have the transit instrument (Fig. 7). A clock is the scale by which differences in right ascensions are measured. The principle of the transit instrument has been retained from Ole Roemer's time up to the present without significant

Fig. 7. The large transit instrument of the Pulkovo Observatory.

changes. Naturally, it has undergone a number of technical modifications and improvements.

The modern transit instrument used in fundamental astrometry has a telescope objective with a diameter of 150–180 mm and a focal length of 150–250 cm. The horizontal axis of the instrument consists of a central hollow cube (sometimes a sphere) and two conical semi-axes ending in two cylindrical pivots. The objective and eyepiece halves of the telescope are also fastened to the cube of the instrument perpendicular to the horizontal axis.

The pivots of the horizontal axis rest in bearings mounted in the first vertical on two stone pillars. These bearings are V-shaped with an angle of 90°, so that ideally each pivot touches its bearing at two points only. In order to avoid wear and tear on the pivots, which causes systematic errors in observation, floating devices which reduce the pressure of the pivots on the bearings to a few kilograms are used. Special screws for the bearings permit altering the position of the horizontal axis by displacement of the bearings.

The eyepiece part of the instrument contains an ocular micrometer with a movable thread and a stationary grid of vertical threads for registering transits. So that the threads of the micrometer may be visible, the field of vision of the instrument is illuminated either by light passing through the semi-axis and reflected from the diagonal mirror in the tube, or by small light bulbs placed in front of the objective. To dim the images of excessively bright stars, special filters are put in the eyepiece or wire screens of various density are put in front of the objective. Recently, photoelectric registration of passages has been made without direct participation of the eye of the observer in sighting stars.

A circular scanner, usually graded in 10′ divisions and used to set the instrument at the required zenithal distance, is seated on the horizontal axis. The instrument has a fastening clamp and a micrometer screw to fix it in elevation and to align it more exactly on a sighted star.

The construction of the instrument allows a 180-degree rotation of the horizontal axis together with the telescope tube around a vertical line. For this purpose, the instrument has a reversing jack by means of which it is rotated from time to time.

During observations, the astronomer is seated in a special-backed chair which allows him to sit with his legs either to the south or to the north (see frontispiece). So that observations may be made more comfortably, the back of the chair can be set at various angles with the horizontal. The chair can be displaced along the meridian up and down a track in the floor and its seat can usually be raised or lowered.

When the instrument is rotated around the horizontal axis, the projection of the line of sight describes a curve on the celestial sphere that differs only slightly from a great circle close to the meridian. An inaccurate orientation of the horizontal axis (its inclination and azimuthal error, or azimuth) and the nonperpendicularity of the line of sight to the axis of the instrument (collimational error or collimation) cause a deviation of the line of sight from the plane of the celestial meridian. Therefore, the observed instant of passage of a heavenly body past the central thread will not be the instant of transit across the celestial meridian, and it will then be necessary to correct these errors. The calculation of all errors misplacing the instant of transit is involved in the theory of the transit instrument.

The right ascension of a culminating star α is equal to the sidereal time S; this last can be obtained from observation if the observed instant of transit T is corrected for an inaccurate orientation of the horizontal axis, the clock error u, and the collimation error ΔT. Consequently,

$$\alpha = T + u + \Delta T.$$

The German astronomer Mayer derived the formula which bears his name and which enables us to calculate ΔT:

$$\alpha = T + u + \underbrace{\sin z_m \sec\delta \cdot k + \cos z_m \sec\delta \cdot i + c\sec\delta}_{\delta},$$

or

$$\alpha = T + u + Kk + Ii + Cc,$$

where $K = \sin z_m \sec\delta$, $I = \cos z_m \sec\delta$, and $C = \sec\delta$ are known as Mayer's coefficients; $z_m = \varphi - \delta$; k and i are the azimuth and the inclination of the horizontal axis and c is the collimation. In this formula, the azimuth is taken as positive if the west end is displaced to the south, the inclination is considered positive if the west end of the axis is higher than the east end, and the collimation is considered positive if the angle between the upper half of the line of sight (toward the objective) and the western half of the horizontal axis is greater than 90°.

In the form in which it is given here, Mayer's formula is valid even for low culminations if the declination is reckoned from 0 to 360° in the direction from the equator to the North Pole and consequently, the zenith distance is reckoned from 0 to 360° in the direction from the zenith to the south point.

Mayer's formula is not exact, but it can be used for small values of i, k, and c.

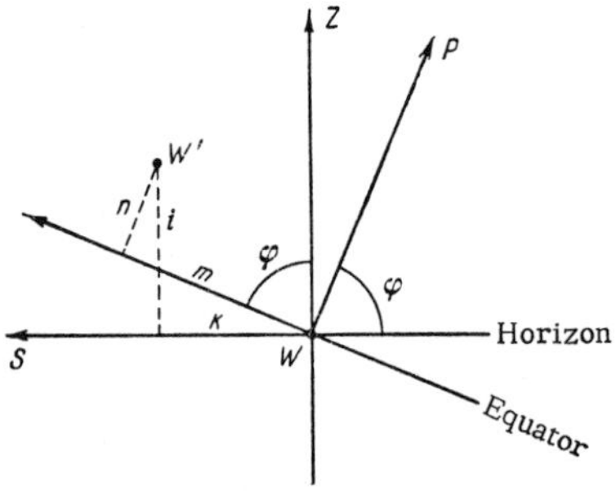

Fig. 8. Transformation from Mayer's formula to Bessel's formula.

Bessel's formula is easily derived from Mayer's formula if we change from horizontal coordinates of the west end of the horizontal axis i and k to the equatorial coordinates $(90° - m)$ and n. Figure 8 shows a portion of the celestial sphere close to the west point, visible from within, with the two systems of coordinates. Because of the small values of i, k, m, and n, we may use the following formulas for changing from one coordinate system to the other:

$$i = m \cos\varphi + n \sin\varphi, \quad k = m \sin\varphi - n \cos\varphi.$$

By substituting these values in Mayer's formula, we obtain Bessel's formula:

$$\alpha = T + u + m + n \tan\delta + c \sec\delta.$$

There is still Hansen's formula

$$\alpha = T + u + i \sec\varphi + n(\tan\delta - \tan\varphi) + c \sec\delta,$$

which is sometimes used in processing observations.

7. Astronomical Clocks

Determination of the instants of star transits across the meridian or the determination of the difference between right ascensions from observations by means of a transit instrument is impossible without the use of an astronomical clock. The adjective "astronomical" indicates that the clock used for this purpose must be extremely accurate; that is, it must possess the greatest stability in its rate.

The basic indicator of the quality of a clock is the uniformity of its rate. The change in the correction of a clock after a 24-hour period is called its diurnal rate, or simply the rate, and is indicated by $\omega = u' - u$, where u and u' are the corrections made for the instants T and $T + 24^h$. If the rate of a clock were constant, then such a clock would run uniformly, and, knowing its rate, we could compute its reading in advance for an arbitrary instant. The diurnal rate of a given clock ω_i can be obtained both from astronomical observations of the corrections of the clock and also from comparison with other clocks whose diurnal rate is known.

The basic property of a clock is the variation in its diurnal rate

$$\delta_\omega = \sqrt{\frac{\sum \Delta\omega_i^2}{n}},$$

where $\Delta\omega_i = \omega_{i+1} - \omega_i$ is the difference in the diurnal rates for consecutive days and n is the number of $\Delta\omega_i$ taken. For an astronomical clock to guarantee the reliability of interpolating corrections in it for the instants of observation, it must have a diurnal variation not exceeding one or two thousandths of a second.

Depending on the periodic regulating process, astronomical clocks can be classed as pendulum, quartz, and atomic or molecular clocks.

Pendulum clocks are based on a pendulum's constant period of oscillation. This characteristic, however, is true only under ideal conditions. The period of oscillation of a pendulum is very sensitive to changes in external conditions, especially temperature, which causes a change in its length and in the elastic properties of the suspending spring. To eliminate this influence, the rod of the pendulum is made of invar or superinvar—alloys having negligible coefficients of linear thermal expansion; also, the change in the length of the rod of the pendulum can be compensated. Furthermore, in order to reduce the influence of temperature changes decisively, the pendulum is placed in a vault at a depth of ten to twenty meters where diurnal changes in temperature are absent and the annual changes are small, not exceeding half a degree.

A variation in pressure causes the density of the medium surrounding the pendulum to change, and this too is reflected in the period of its oscillation. Therefore, the pendulum is enclosed in a hermetically sealed copper cylinder. In such a cylinder, a constant low pressure of the order of twenty millimeters is maintained. The length of the pendulum is usually about one meter, so that the period of one oscillation is equal to two seconds.

The most perfect pendulum clocks are those with a so-called free pendulum. The free pendulum is used under the conditions described above so that its oscillations will be as nearly constant as possible.

The free pendulum does not move the mechanism of the clock and is in contact with the other parts of the mechanism only at the instant of winding. In order that the oscillations may not be damped, especially by friction with the spring suspension of the pendulum, the winding consists of a light push once every thirty seconds. By means of an electric connection, a second pendulum subject to ordinary room conditions is forced into oscillations synchronized with the main pendulum. When the pendulum of the secondary clock passes through its vertical position, it closes the seconds contacts. Half-minute contacts are more exact because they agree with the definite positions of the free pendulum.

The accuracy of pendulum clocks is relatively high: for example, the variation in the rate of Shortt's clock is usually of the order of $\pm 0^s.001$-$0^s.002$—and it works continuously for several years. Fedchenko's pendulum clock, which was constructed in the Soviet Union a few years ago, has a variation in the rate not exceeding $\pm 0^s.0003$. Until the 1930's, pendulum clocks were the only ones used in meridian astrometry.

In 1927, Morrison and Horton constructed a quartz clock which was first introduced into astronomical work at the Potsdam Observatory. It is a complicated radio technical device and is a most accurate timekeeper. The variation in the diurnal rate of a good quartz clock amounts to $\pm 0^s.0001$-$0^s.0002$.

The construction of a quartz clock is based on the piezoelectric property of a quartz crystal. If a quartz disk is deformed, electric charges appear on its surface and, conversely, if it is placed in a varying electric field, it will undergo elastic vibrations.

A quartz clock consists of the following basic parts: a quartz generator, which produces a stable frequency, a divider and an amplifier to transform this frequency, and a clock mechanism.

The quartz generator produces a variable alternating current of constant frequency corresponding to the natural frequency of oscillation of the quartz disk or ring. Since the natural frequency of the quartz depends on the temperature, the quartz disk is regulated by a thermostat, where the temperature is kept constant with an accuracy of $\pm 0^\circ.01$ or better. In order to eliminate the influence of the change in atmospheric pressure and humidity, the quartz is kept in an evacuated sealed glass container.

The frequency generator, in whose electric field the quartz is located, is tuned so that its frequency of oscillation will coincide with the natural frequency of the mechanical vibrations of the quartz, the amplitude of which will be maximum upon coincidence with the frequency of the imposed electric field. The electric vibrations appearing on the facing of the quartz as a result of the piezoelectric effect are standard vibrations and serve as a measurement of time. The role of the generator consists in maintaining the mechanical vibrations of the quartz. The natural frequency of vibration of the quartz used in quartz clocks depends on its dimensions and shape and is between 50,000-1,000,000 cycles per second.

So that the stable high-frequency current may be put to use, it is amplified and its frequency is decreased by means of a divider. The stable low-frequency current obtained (of the order of 1,000 cycles per second) feeds a synchronous motor, which rotates with a constant velocity if the frequency of the current feeding it is constant.

Specially designed mechanical reducers of the rotation velocity impart driving motion to the hands of the clock. The seconds contact is closed by a disk which makes one rotation per second. This mechanism does, however, cause some random scattering of the seconds contacts. Therefore, quartz clocks are sometimes made without a synchronous motor and mechanical contacts. By changing the frequency generated by the quartz, short successive pulses are obtained for every other second by means of an electric circuit.

Quartz clocks, while the best of contemporary clocks, are not free from certain defects. Any disorder in the electrical system—for example, a failure of one of the numerous radio tubes—stops the clock. It is possible to prolong the period of operation of a clock by feeding several dividers, amplifiers, and clock mechanisms from one quartz generator. Then, only disturbances in the quartz generator itself will cause the clock to stop. Gradual and abrupt changes in the natural frequency of the quartz constitute another defect. The gradual change in this frequency—the aging of the quartz—ordinarily disappears after one or two years of use of the clock. To control the abrupt changes, which usually take place no oftener than once in a year, one must have at least three quartz clocks.

Recently, scientists have developed atomic and molecular clocks in which the natural frequency of vibration of a molecule or atom of certain substances is used as a frequency standard. Atomic and molecular clocks now serve only as frequency standards since they can keep time for only a relatively short period; they are used in a single system with quartz clocks, checking and correcting the rate of the latter.

8. Chronographs

A chronograph is a device for reading accurately the instant of a particular event. A chronograph registers the instant of closing or opening of a contact in an electric circuit. In astrometric work, when determining coordinates, astronomers use a chronograph for registering the instants of passage of a heavenly body past particular points in the field of vision. From the standpoint of construction, chronographs are divided into (a) writing or punching, (b) printing, and (c) cylindrical devices. The third kind, which in practice is chiefly used in time services for continuous comparison of several clocks, is rarely used in astronomical observations.

The writing chronograph includes a tape-transport device and a mechanism for recording on the tape the instants of closing or opening of contacts. This mechanism has writing pens or puncturing needles. A paper tape is set in uniform motion by a clockwork driven by a spring or by weights. The pens are in contact with the moving paper tape and trace a line on it. When the contacts connected to the chronograph through a relay are closed, the circuit of the electromagnet, which attracts a lever with a pen, is closed; this causes a break in the line on the tape. When the contacts open, the pen returns to its initial position. Because of the relay, the currents passing through the contacts of the micrometer and the contacts of the astronomical clock are weak. This is very important; if the contacts are allowed to burn, timing errors will result.

The seconds and half-minute contacts of an astronomical clock are connected to the circuit of the electromagnet of one of the pens. The contacts of a micrometer are connected to the other pen. The recording on the tape takes the form shown in Fig. 9.

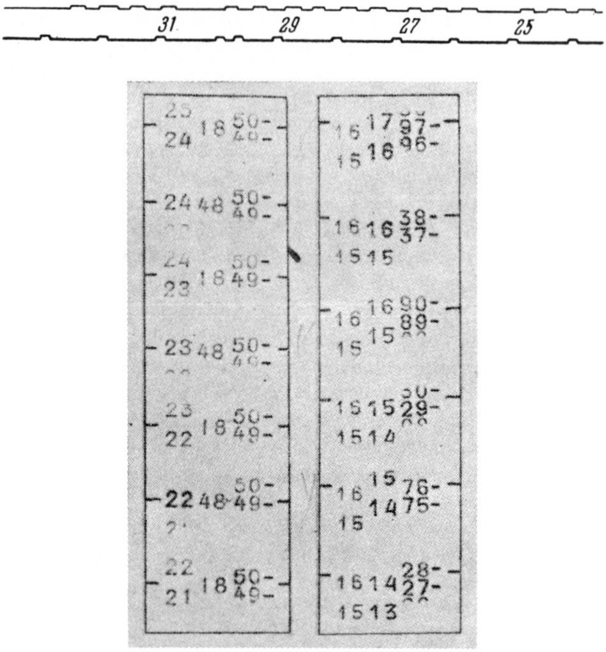

Fig. 9. Tapes of writing and printing chronographs.

In order to determine the instant of the contact made by the micrometer pen, it is necessary to measure on the tape the distance from the contact to the nearest seconds contact. This is accomplished with a special measuring device. The mean square error of registration of an instant on a writing chronograph is close to $\pm 0^s.01 - 0^s.02$. The measurement of the tape is a laborious process, and for that reason, at the present time, writing chronographs are being replaced by printing chronographs.

Let us describe the printing chronograph developed by the Scientific Research Institute of the Clock Industry by order of the Moscow State University (Fig. 10). The successful construction and the practicality of these chronographs made possible their wide distribution. They are used throughout the Soviet Union in connection with meridian observations.

The basic feature of the printing chronograph consists of three disks connected with each other by a planetary transmission and set in continuous rotation through reduction gearing by a synchronous motor. The minute and seconds disks have circular dials numbered from 0 to 59. The third disk, indicating hundredths of a second, is numbered from 00 to 99 and has opposite each number a short line to aid interpolation to the thousandths of a second from the position of the imprint of the

Fig. 10. A printing chronograph.

immovable index. This disk makes one rotation per second, the seconds disk makes one rotation per minute, and the minute disk makes one rotation per hour. Uniformity of rotation of the disks is made possible by the stability of the frequency of the current feeding the motor.

This current, with a frequency of one hundred cycles per second, is generated by a special quartz generator which is part of the chronograph. The principle of construction of the chronograph quartz generator is analogous to the principle of construction of the generator in quartz clocks. The difference lies in the cruder thermostat apparatus of the quartz disk and in the simplified design of the frequency amplifier and divider. The motor of the printing chronograph can also be fed from the quartz clocks by a current of stable frequency. Naturally, the operating accuracy of the chronograph is then increased. The quartz generator produces a frequency such that the chronograph runs on mean time. By a slight modification of the quartz disk, a mean-time chronograph can be converted into a sidereal chronograph, which is more suitable for processing observation data. The quartz generator and chronograph can be fed only from alternating current of industrial frequency.

Under the rotating disks of the printing chronograph there is a typewriter ribbon and under it a paper tape. Beneath that is a small impact hammer. When an impulse is received at the entrance relay of the chronograph, the circuit of the impact mechanism is closed and the hammer sharply and quickly (so as not to disturb the rotation of the disks) strikes the paper, pressing it and the typewriter ribbon against the numbers of the disks which at that moment are opposite the hammer. As a result, the paper tape gets the imprint of the minutes, seconds, and hundredths of a second and also the imprint of the motionless index; the index imprint permits visual estimate of thousandths of a second from its position relative to the hundredths (see Fig. 9).

After the striking of the hammer, the tape-transport mechanism goes into action and the paper tape moves on for the next imprint. A special switch allows one to obtain the imprints only at the beginning of an impulse, only at the end, or both together. The chronograph can register pulses with duration as short as $0^s.08$. The minimum permissible interval between impulses is $0^s.017$. The mean square error of the instants registered on a printing chronograph is approximately $\pm 0^s.003$.

A quartz generator works steadily through the course of a night of observations; stability of operation of the chronograph is ordinarily attained after the generator has been working one or two hours. Therefore, it is not necessary to print the impulses of the contacts of an astronomical clock often. To control the speed of the chronograph, this should be done several times during a night of observation.

When one is working with a printing chronograph, it is possible to make an imprint with a key and to let the tape run without an imprint; this is convenient for separating the imprints relative to the different stars. The chronograph is equipped with a remote-control attachment which allows one to control it from the pavilion if it is set up outside or at a considerable distance from the instrument.

There are chronographs in which the printing device is replaced with a photocamera. A photoprint of the position of the rotating disks relative to the stationary index is made when a neon light flashes. When the contact is closed, a condenser included in the same circuit as the neon light discharges. This system acts almost without inertia.

9. The Vertical Circle and Measurement of Zenithal Distances

In order to determine declinations from meridian observations, it is necessary to be able to measure zenithal distances of stars. This is possible with the help of two instruments: the vertical circle and the meridian circle.

The idea of the vertical circle and its first use in fundamental astrometry belong to the founder of the Pulkovo Observatory, V. Struve. Its prototype was the geodesic universal instrument. In contrast with the latter, the new instrument has only one vertical circle as a basic circle, and hence the name.

We shall give a short description of the construction of the vertical circle (Fig. 11). The main column of the instrument stands on three legs on a large base. Two legs stand in the direction of the prime vertical so that it is possible to correct the position of the vertical axis in the plane of the line of sight—that is, in the plane of the meridian—by means of a screw on the third leg. Inside the column is the vertical axis of the instrument, which is cone-shaped toward the top but which is fixed in position by a cylindrical stand toward the bottom. By means of levers and weights, a floating force is applied at the lower end of the axis for making rotation easier.

A large perpendicular bar contains the bearings for the horizontal axis and is placed at the upper end of the vertical axis. A circular or a rectangular frame with microscopes is fastened to it. The horizontal axis of the instrument, with a telescope tube at one end and a counterweight at the other, rests in a V-shaped bearing. On the horizontal axis near the tube there is an accurately graduated circle which serves for measuring zenithal distances. The horizontal axis has weight-reducing counterpoises, a clamp, and a micrometric vertical adjustment.

The ocular micrometer of the telescope has a stationary grid of vertical threads and a movable horizontal thread for bringing the tube to bear on a star.

On the frame or drum with the microscopes in the plane of the line of sight, there is a level for measuring the change in the position of the frame of the microscopes relative to the horizon when the instrument is rotated through 180° around the vertical axis. There is a striding level

Fig. 11. The vertical circle of the Pulkovo Observatory.

for controlling the leveling of the instrument and the inclination of the horizontal axis.

Sometimes, at the end of the horizontal axis, a finder-circle is placed beside the counterweight and a steering control for rotating the instrument up and down. On the vertical axis, there is a coarse-graduated circle, used for placing the instrument in the proper azimuth. At the upper (sometimes at the lower) portion of the column are adjustable end supports, regulated to set the instrument exactly on the meridian. This makes it easier and quicker to rotate the instrument through 180° around the vertical axis from one position to another.

The theory of the vertical circle coincides completely with the theory of the universal instrument described in courses in practical astronomy as far as the absolute measurement of zenithal distances is concerned. Therefore, we shall here discuss only briefly the facts necessary for further presentation.

The principle of determining zenithal distances is illustrated in Fig. 12, in which the frame with the microscopes is schematically represented in the form of an index with a level.

Suppose that I is the position of the instrument at the initial sighting of an object with zenithal distance z. The reading with respect to the index J_{I} is M_{I}; we shall assume the direction of graduation to be such that $z = M_{\mathrm{I}} - M_z$, where M_z is the zenith point, the reading on the

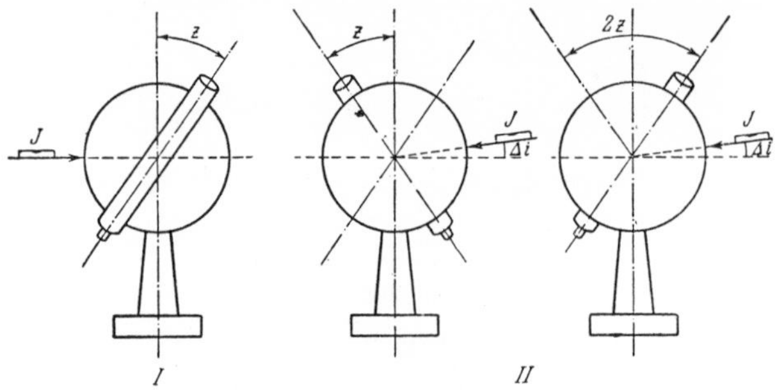

Fig. 12. Determination of zenithal distances with the vertical circle.

calibrated circle of the instrument when the tube is directed toward the zenith. We shall consider refraction to be negligible. Let us turn the instrument through 180° around the vertical axis into the second position. Now, in order to see the object on the same threads, we need to turn the telescope around the horizontal axis, and the reading on the circle becomes M_{II}.

If the position of the frame with the microscopes or, what is the same thing, that of the index relative to the horizontal plane remained the same after the rotation of the instrument, this would allow the point at the zenith to remain constant. However, because of the inclination of the vertical axis in the plane of sighting, because of the change in this inclination in going to the second position, and because of a possible rotation of the frame with the microscopes as a whole, the position of the index relative to the horizontal line will change and it will occupy the position J_{II}, which differs from the position J_I by an amount Δi. The point at the zenith will change by the same amount Δi and the zenithal distance of the object will be $z = M_z - \Delta i - M_{II}$.

Thus, for the two positions of the instrument, z will have an average value $z = \frac{1}{2}[M_I - M_{II} - \Delta i]$. Consequently, there is no necessity to determine the point of the zenith, and we need only to know how to compute the change in it when we go from one position of the instrument to the other. All of this can be computed from the readings of an alidade level attached to the frame of the microscopes and whose axis is parallel to the plane of the line of sight. If a_1 and b_1 are the readings at the ends of the bubble at the first position of the instrument, the inclination of the axis of the level with respect to the horizontal line will be

$$i_1 = \frac{\alpha''}{2}[(a_1 + b_1) - (a_0 + b_0)],$$

where α'' is the value of the division of the level, and a_0 and b_0 are the readings when the level is in a horizontal position. After the instrument has been rotated through 180°, the inclination of the axis of the level will be

$$i_2'' = \frac{\alpha}{2}[(a_2 + b_2) - (a_0 + b_0)],$$

where a_2 and b_2 are the readings of the bubble in the second position.

A change in the inclination of the axis of the level

$$\Delta i = i_2 - i_1 = \frac{\alpha''}{2} [(a_2 + b_2) - (a_1 + b_1)] \quad (1)$$

corresponds to a change in the zenith point when we go from the first position of the instrument to the second. The sign of the correction is determined by the directions of the graduations of the level and of the graduated circle, and by the sequence of positions. As can be seen from formula (1), the zero point of the level is excluded and, consequently, it is necessary that it should not change its position relative to the instrument during the time of observation of one star. The security with which the level is fastened makes this entirely possible.

Another valuable feature of the vertical circle should be mentioned. As a means of making measurements more accurate, a star is sighted by a movable horizontal thread of the ocular micrometer, and not by the entire instrument. It is then possible to make several sightings of a star with one reading of the graduated circle. The readings of the micrometer $m - m_0$ are added to or substracted from the readings of the graduated circle, depending on the directions of the graduations of the circle and the micrometer. The value m_0 is the zero point of the micrometer; that is, the reading corresponding to the center of the field of vision in which the sighting is done. The zero point, taken provisionally for a given series of observations, fixes the relative position of the readings of the micrometer and the graduated circle. Since z is the difference in the readings, the zero point of the micrometer is eliminated, and since observations of a single star take only a brief instant of time, its variations can be neglected.

Thus, it is only necessary for the zero points of the level and of the micrometer to be constant during the time of the observation of a single star. Their variations in the course of an evening have no influence on the measured zenithal distances.

A serious imperfection of the vertical circle is its asymmetry, which sometimes apparently causes great systematic errors in observations of declinations.

An interesting construction of a photographic vertical circle is being developed at the Pulkovo Observatory following an idea of M. S. Zverev. The chief characteristic of this instrument is the use of the meniscus telescope of D. D. Maksutov, which makes it possible to construct a symmetrical and compact instrument. Instead of sighting by means of the thread of the ocular micrometer, the new method is to photograph passages of the stars for two positions of the instrument and to measure the distances between the tracings of the images of the stars on the photoplate. The readings of the graduated circles will also be done by the photographic method.

10. The Meridian Circle and Its Uses

The meridian circle (Fig. 13) is a transit instrument equipped with a precisely graduated circle for measuring angles in the vertical plane. It is used for determining both right ascensions and declinations. The graduated circles, of which there are ordinarily two, are placed on the horizontal axis in symmetrical positions on either side of the telescope. One of them is the finder-circle. Calibrated microscopes with graduated heads are placed on supporting pillars of the instrument for reading the graduated circle. Usually, four microscopes are used for reading, sometimes two or six. In addition to a screw for right ascensions, the ocular micrometer of a meridian circle also has a screw for declinations.

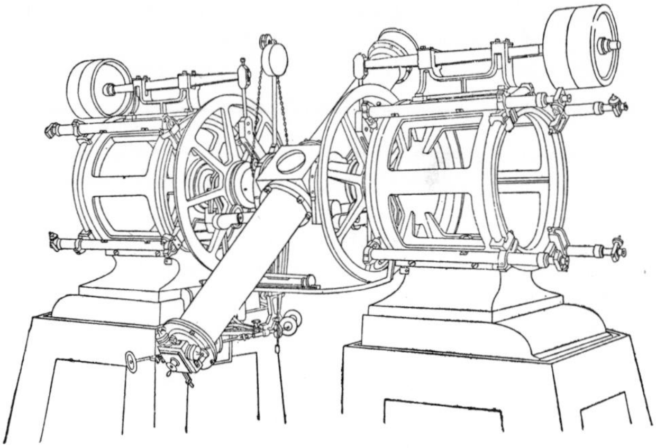

Fig. 13. Diagram of a meridian circle.

During observation of a star moving along a diurnal parallel in the field of vision of a meridian circle, it is possible to register by the clock the instant of passage of the star across the meridian of the instrument and to measure the arc in the plane of the meridian. Since one cannot reset the meridian circle for the observation of every star, as one can for the vertical circle, it is then necessary to determine the zenith point. The surface of mercury in a basin under the instrument serves for a horizon.

The data obtained from observations with a meridian circle make it possible to determine the equatorial coordinates of heavenly bodies. For this, the clock serves as a scale for measuring the right ascensions and the graduated circle for measuring the declinations. The registering micrometer and the chronograph are reading devices for registering the instants by the clock, and the graduated heads of the microscopes are for reading the graduated circle. From the observations of a single passage of a star across the meridian, it is possible to determine both coordinates with the meridian circle. These combined observations, obtained only by means of meridian circles, have especially wide application in relative determinations of stellar coordinates.

The theory and practice of observations of the moments of stellar crossing of the meridian on a transit instrument are no different from those on the meridian circle; therefore, the accuracy of determination on both instruments is the same. Determination of the declination on the vertical circle, however, is somewhat more exact because of the elimination of the zenith point on every star and because of the double reading of the circle. The accuracy of observation of the coordinates obtainable from agreement between the different determinations on a given instrument is characterized by the following numbers. The mean square error of a single determination of right ascensions on a transit instrument or a meridian circle is $\pm 0^s.02 \sec \delta$; the accuracy of determining declinations on a meridian circle $\pm 0''.45$ and on the vertical circle $\pm 0''.35$.

The question as to what fundamental astrometric instruments or what observational methods are the most expedient sometimes depends on economic considerations and also on the personal preferences of the astronomers. One thing is clear: to obtain a significant increase in the accuracy of determining coordinates, we must introduce essentially new instruments and methods.

The simplicity of construction and theory of the meridian circle and its universal use have made it, up to now, the basic instrument for determining the coordinates of celestial bodies, especially the stars. At the present time there are approximately one hundred meridian circles in the world and less than half of them are used in continuous observation. In the USSR, active observations are being carried out on all nine existing instruments.

11. Construction of Foundation and Pavilion of a Meridian Circle

For the foundation of a meridian circle, as for any astrometric instrument, stability and solidity are required. The foundation must not be deformed either by its setting or by structural changes.

The foundation of the instrument (Fig. 14) is ordinarily a massive brick block laid on the ground with its base lower than the level at which the soil freezes (at middle latitudes, about 1.5 m). Sometimes, channels are left in the body of the foundation to insure a more uniform distribution of its temperature.

On the upper surface of the foundation there is a monolithic stone slab on which the pillars of the instrument itself are placed for stabilizing the bearings and the heads of the calibrated microscopes. The pillars are made of single stone blocks, usually of granite, or of laid brick. Some instruments have metal pillars. Investigations have shown that single-stone pillars are the most stable, and therefore the azimuth and the inclination of the horizontal axis change the least. On the outside, the pillars of the instrument are usually protected from the influence of sharp temperature changes by the use of wooden sheathing.

The foundation and the pillars of the instrument must be laid down independently of the foundation and the walls of the pavilion, and must not come in direct contact with the floor of the working area lest vibrations propagated along the surface of the earth or vibrations caused by the observer be transmitted to the instrument. The pillars for the collimators and the other control devices are set up either independently or on a common foundation with the instrument. As a rule, they are of brick; the stability requirements are not so high with them as with the pillars of the meridian circle.

The pavilion of the astrometric instrument, in addition to protecting it from its surroundings, must protect it from the heat of the sun. Also,

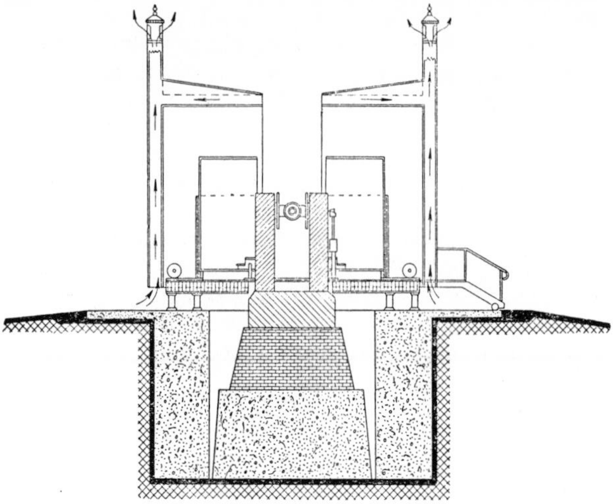

Fig. 14. Cross section of pavilion and foundation of a meridian circle.

the construction of the pavilion must allow rapid equalization of the temperature of the instrument with the temperature of the outside air both at the beginning and during the course of observations. All of this determines the arrangement of the walls and the sliding shutters of the pavilion.

The walls of a good pavilion are made hollow, with an inner air space ventilated either by a natural or a forced draft. Also, the inner surface of the wall will have a cover of felt, plastic foam, or some other insulating material. A metallic pavilion is sometimes covered on the outside by a jalousie-type grill of blocks of wood, which protects the pavilion from direct sunlight but allows access of air to the metallic covering. It is always desirable to plant trees, bushes, etc., around the pavilion to give it shade.

The pavilion must have a slit in the roof that may be opened along the meridian. The pavilion often consists of two separated halves. The slit should be no less than two meters wide so that the instrument may quickly acquire the temperature of the outside air. A wide slit also minimizes the so-called "room" refraction, caused basically by the difference in temperature inside and outside the pavilion.

Investigations have shown that the narrow slits (0.5–1.0 m) used in the past and the location of the instrument in large pavilions do not afford the best conditions for observation; the pavilions so constructed must now be considered obsolete. Therefore, astronomers are striving to transfer old instruments from large pavilions, and set them up, along with new instruments, in small pavilions with wide openings and good ventilation (Fig. 15). In this connection, it is not good to place the instrument directly in the open air because of the effect of the wind.

As a protection from the wind at the time of observations, especially with a wide opening, cloth curtains of adjustable height are set up along the slit on the north and south sides. The south curtain is also used to protect the meridian circle from the rays of the sun when the sun is

Fig. 15. One of the pavilions of Pulkovo Observatory.

being observed. In the day-time, observations are made through an opening cut in the curtain. Sometimes, in order to protect the instrument from dust within the pavilion, a removable canvas cover is placed over it.

12. The General Theory of Meridian Instruments

The formulas used in practice for computing equatorial coordinates are approximations introduced on the assumption that the errors of orientation of the axes and of the optical line of sight of the instrument are small with respect to each other and with respect to the plane of the horizon. Therefore, it is necessary to know the admissible limits of these errors in order to correct the setup of the instrument as the parameters change. An estimate of these limits is possible if we derive the rigorous formulas of meridian instruments.

We begin by deriving a rigorous formula for a transit instrument. In Fig. 16, suppose that W' represents the projection of the western end of the horizontal axis on the celestial sphere. As a result of errors in setting up the instrument, i and k (or n and m), and the collimation c, a star σ passes across the observation thread of the instrument outside the meridian, and the recorded instant T requires correction ΔT.

Let us examine the triangle $W'Z\sigma$. Its sides are: $W'Z = 90° - i$, $Z\sigma = z$, $W'\sigma = 90° + c$. The angle at Z is equal to $(90° - k - A)$, where A is the azimuth of the star. In Fig. 16, k, the azimuth of the horizontal axis, is negative. From the cosine formula (when three sides of the triangle are given) we have

$$\cos(90° + c) = \cos(90° - i)\cos z + \\ + \sin(90° - i)\sin z \cos(90° - k - A),$$

or

$$-\sin c = \sin i \cos z + \cos i \sin z \sin k \cos A + \\ + \cos i \sin z \cos k \sin A.$$

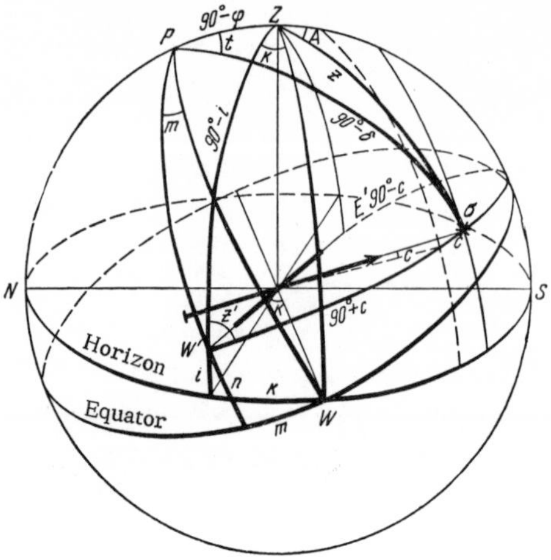

Fig. 16. The general theory of meridian instruments.

Using the three standard formulas for the spherical triangle $PZ\sigma$, namely,

$$\begin{aligned}\cos z &= \sin\varphi\sin\delta + \cos\varphi\cos\delta\cos t,\\ \sin z\cos A &= -\cos\varphi\sin\delta + \sin\varphi\cos\delta\cos t,\\ \sin z\sin A &= \sin t\cos\delta,\end{aligned}$$

we can eliminate the horizontal coordinates A and z of the star:

$$\begin{aligned}-\sin c = \sin i\,&(\sin\varphi\sin\delta + \cos\varphi\cos\delta\cos t) +\\ &+ \cos i\sin k\,(-\cos\varphi\sin\delta +\\ &\quad + \sin\varphi\cos\delta\cos t) + \cos i\cos k\sin t\cos\delta,\end{aligned}$$

or, by letting $z_m = (\varphi - \delta)$, we obtain the exact formula for a transit instrument

$$\begin{aligned}\sin c + \sin i&\left(\cos z_m - 2\cos\varphi\cos\delta\sin^2\frac{t}{2}\right) +\\ &+ \cos i\sin k\left(\sin z_m - 2\sin\varphi\cos\delta\sin^2\frac{t}{2}\right) +\\ &\qquad + \cos i\cos k\cos\delta\sin t = 0.\end{aligned}$$

We note that this relation is accurate for any vertical.

The errors of the mounting of the horizontal axis and of the collimation (i, k, and c) must be held within certain limits so that the values of the reduction to the meridian, as computed by Mayer's formula

$$\Delta T = i\cos(\varphi - \delta)\sec\delta + k\sin(\varphi - \delta)\sec\delta + c\sec\delta,$$

differ from the value computed from the exact formula by a negligible amount only. To estimate these, we transform the exact formula for the transit instrument. Since the values of i, k, c and t are small, we replace the trigonometric functions of these arguments by their series expansions and limit the accuracy to the third-order terms:

$$\left(c-\frac{c^3}{6}\right)+\left(i-\frac{i^3}{6}\right)\left[\cos z_m-2\cos\varphi\cos\delta\frac{t^2}{4}\right]+$$
$$+\left(1-\frac{i^2}{2}\right)\left(k-\frac{k^3}{6}\right)\left[\sin z_m-2\sin\varphi\cos\delta\frac{t^2}{4}\right]+$$
$$+\left(1-\frac{i^2}{2}\right)\left(1-\frac{k^2}{2}\right)\left(t-\frac{t^3}{6}\right)\cos\delta=0.$$

We expand the expressions in parentheses, keeping the third-order terms, and transpose the term containing the first power of t equal to ΔT. Assuming i, k, c, and t expressed in seconds of time, we obtain

$$-t=i\cos z_m\sec\delta+k\sin z_m\sec\delta+c\sec\delta-$$
$$-\frac{15^2}{206\,265^2}\left[\frac{i^3}{6}\cos z_m\sec\delta+\frac{k^3}{6}\sin z_m\sec\delta+\right.$$
$$+\frac{c^3}{6}\sec\delta+\frac{ki^2}{2}\sin z_m\sec\delta+t\left(\frac{1}{2}i^2+\frac{1}{2}k^2\right)+$$
$$\left.+t^2\left(\frac{i}{2}\cos\varphi+\frac{k}{2}\sin\varphi\right)+\frac{t^3}{6}\right]. \qquad (2)$$

The first three terms constitute Mayer's formula. Clearly, it is accurate up to third-order terms, the influence of which rises sharply for high declinations.

It is necessary to estimate the influence of the third-order terms. To do this, we need to assign certain values to i, k, and c. The inclination is well defined by the striding level and never exceeds 1^s. The azimuth of the horizontal axis does not ordinarily exceed 1-2^s. The collimation is easily regulated by moving the grid of threads in the ocular micrometer; it is always small. However, we should note that observations are often made on the side threads, that is, not in the center of the field of vision. Therefore, by "collimation" we mean the sum of the collimation and the distance of the side thread from the middle thread. This last may attain the significant value of 10-20^s.

Let us compute the value of the third-order terms for stars with declinations equal to +80° and +85°, with $\varphi = +55°$. Clearly, on the right side of equation (2), t can be replaced according to the formula

$$t=-i\cos z_m\sec\delta-k\sin z_m\sec\delta-c\sec\delta.$$

For i, k, and c, we take two sets of values: (1) $i = 3^s$, $k = -3^s$, $c = 30^s$ and (2) $i = 1^s$, $k = -1^s$, $c = 10^s$. The azimuth is taken as negative because in this case its influence is combined with the influence of the inclination and the collimation. Under these conditions, the values of the third-order terms will be: for $\delta = 80°$, (1) $-0^s.001$, (2) $-0^s.000$; and for $\delta = 85°$, (1) $-0^s.054$, (2) $-0^s.002$.

Thus, third-order terms should be taken into account when observations are made on the far side threads for declinations greater than 80 degrees. If observations are made on the central threads ($c < 10^s$), the third-order terms need not be considered.

The problem of determining declinations amounts to determining zenithal distances from observations. Let us examine the effect of errors made in setting up the instrument in the case of measurement of the zenithal distances by means of a vertical circle. Among these errors, we include the inclination i and azimuth k of the horizontal axis, the inclination of the vertical axis in the plane of the line of sight and the collimation c. The value of i is caused by the inclination of the vertical axis in the plane perpendicular to the plane of the line of sight, and by the nonperpendicularity of the horizontal and vertical axes.

Calculation of the influence of the inclination of the vertical axis in the plane of line of sight on the reading of the graduated circle is done by using the level attached to the frame with the microscopes (see Section 9). Therefore, it is necessary to estimate the influence of the remaining errors.

Let us examine (Fig. 16) the parallactic triangle $PZ\sigma$ and the triangle $Z\sigma W'$, the sides of which are $ZW' = 90° - i$ and $\sigma W' = 90° + c$, and whose angle at W' is z'. According to the cosine formula, we have for the triangle $PZ\sigma$

$$\cos z = \sin\varphi \sin\delta + \cos\varphi \cos\delta \cos t = \\ = \cos z_m - 2\cos\varphi \cos\delta \sin^2\frac{t}{2},$$

and for the triangle $Z\sigma W'$

$$\cos z = -\sin i \sin c + \cos i \cos c \cos z'.$$

Equating these expressions and letting $\Delta z = z_m - z'$, we obtain

$$\cos z_m = -\sin i \sin c + \cos i \cos c \cos(z_m - \Delta z) + \\ + 2\cos\varphi \cos\delta \sin^2\frac{t}{2}.$$

Let us expand the functions of the small angles in a series and confine ourselves to the first two powers of the quantities i, c, t, and Δz. We then have

$$\Delta z = \left[ic + \frac{i^2 + c^2}{2}\cos z_m - \frac{t^2}{2}\cos\delta\cos\varphi\right]\csc z_m.$$

After several rather complicated transformations, we obtain the final formula

$$\Delta z'' = \frac{15^2}{206\,265}\left(\frac{t^2}{2}\right)\sin 2\delta + \frac{15^2}{206\,265}\left[\frac{1}{2}(i^2 - k^2)\sin z_m \cos z_m + \\ + ic\sin z_m - kc\cos z_m - ki\cos^2 z_m\right]. \qquad (3)$$

A complete derivation can be found in the book by B. V. Numerov, "The Theory of the Universal Instrument." Formula (3) is suitable both for the vertical circle and for the meridian instrument. The first term is called the correction for curvature of the parallel and is always taken into account when observations are being reduced. The calculation of this correction will be discussed in Section 47.

Let us estimate the influence of the errors i, k, and c on the observed zenithal distances. The terms with the collimation are the largest terms in the square brackets since the observations are usually carried out on the side threads.

Let us compute the value of the quantity in the square brackets. It attains its maximum value for stars observed at the zenith. Assuming

$i = 1^s$, $k = 1^s$, $c = 30^s$, we obtain $-0''.03$, that is, a quantity that can be neglected for observations on a meridian circle.

In the case of the vertical circle, when observations are carried out in two positions of the instrument, an error in the zenithal distance takes the form

$$\Delta z = \frac{1}{2}\left(\frac{i_1^2 + i_2^2}{2} - \frac{k_1^2 + k_2^2}{2}\right)\sin z_m \cos z_m + \\ + \frac{1}{2} c\,(i_1 - i_2)\sin z_m - \frac{1}{2} c\,(k_1 - k_2)\cos z_m - \\ - \frac{1}{2}(i_1 k_1 + i_2 k_2)\cos^2 z_m.$$

In this formula, the indices 1 and 2 refer to observations in the first and second positions of the instrument. As usual, we assume that $c_1 = -c_2$. If we take $c = 40^s$, then, for its influence not to exceed $\pm 0''.02$, it is necessary that $i_1 - i_2$ and $k_1 - k_2$ should not exceed $0'.2$. The quantities i and k themselves can attain a value of $2'$. They are checked by observing the transits of stars with different declinations and by using the readings of the striding level.

CHAPTER III

The Horizontal Axis

13. The Structure of the Horizontal Axis of the Meridian Instrument

The horizontal axis, together with the telescope tube, is the central feature of the meridian instrument.

In the middle portion of the horizontal axis, there is a cube (see Fig. 13), to which are fastened the two semi-axes, as well as the objective and ocular halves of the tube. However, with the horizontal axis thus composed of separate parts, some variation in the relative positions of the individual parts of the instrument is possible. Therefore, in the latest models of the instrument, the horizontal axes are made from a single piece of metal and the cube in the central part of the axis is replaced with a sphere with a harder surface.

The ends of the horizontal axis, which rest upon V-shaped bearings placed on the pillars, have a cylindrical shape and are called pivots. The pivots are made from the very hardest alloys of steel and they are either pressed into the ends of the horizontal axis or they are fitted on them in the form of hollow cylinders. The finishing of the surface of the pivots is carried out after they are put on the horizontal axis. The pivots must be coaxial circular cylinders with the same diameters.

In present-day instruments, the horizontal axis has a length of about one meter and the diameters of the pivots are from 40 to 100 millimeters. A greater length of the horizontal axis is preferable: this decreases the effect of the errors in the pivots and increases the stability of the instrument. The wear and tear on the pivots is decreased when their diameters are larger.

The horizontal axis with its pivots lies in bearings made in the form of plates with V-shaped grooves. Two flanges are fastened to the side grooves of the bearings at a 90-degree angle. The surface of the flange turned toward the pivot is part of a cylinder whose generatrix is perpendicular to the generatrix of the pivot; as a result, the pivot ideally touches the flange at one point. In practice, the pivot rubs a certain area on the flange; hence, to avoid a great amount of wear and tear on the pivots, the flanges are made of a softer metal (ordinarily brass) or even of wood (boxwood). The flanges of the bearing must be fastened in such a way that they can be moved along the horizontal axis. This makes it possible, as the pivots become worn, to change their working section; that is, the section where the pivot touches the bearing.

The bearings of the instrument are placed on metal plates fastened to the upper surface of the stone pillars of the instrument. Such a setup is more stable and is therefore preferable to the arrangement of the bearings whereby the plates are fastened to the inner surface of the pillars. To adjust the position of the horizontal axis with respect to azimuth and to correct its inclination, the bearings (Fig. 17) can be moved by the micrometric screws A, and can be fastened by the clamping screws B. Ordinarily, one bearing is moved up and down, which is most easily done by means of two movable wedges (Fig. 17a), and the other is moved horizontally along the guiding slide (Fig. 17b).

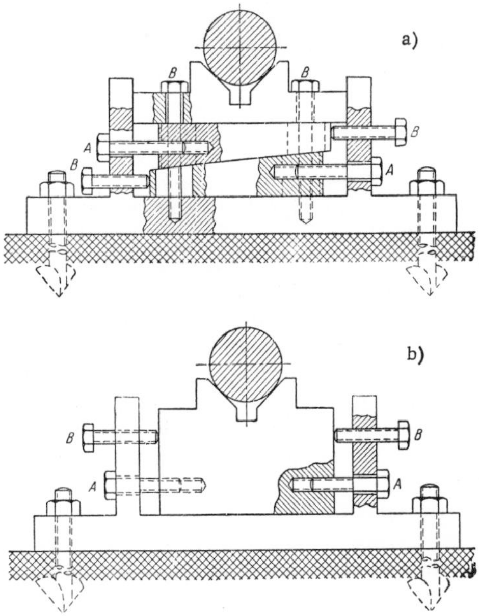

Fig. 17. Adjustment of the horizontal axis for elevation and azimuth.

To fix the position of the horizontal axis with the graduated circle relative to the frame with the microscopes, axial supports are set up. Each of them can be either fastened or free. In the latter case, the support presses by means of a spring on the end of the horizontal axis. When the instrument is in its operating position, the support placed on the side of the microscopes used for reading the instrument is fastened. The opposite support is free and, by means of a strong spring, presses on the end of the horizontal axis. Therefore, the instrument has no free play along the horizontal axis and the distance from the limb of the graduated circle to the reading microscopes is kept constant.

The axis of the instrument is hollow. This makes it possible to illuminate the field of vision by means of a light placed in the first vertical and a diagonal mirror in the cube; it also makes possible a large number of investigations of errors in the instrument.

On those faces of the cube of the instrument that are not fastened to the semi-axes and the telescope (or on the corresponding parts of the central sphere), there are openings with easily removable lids making it possible to direct the collimators toward one another when the telescope is in a vertical position.

In addition to the graduated circles, the horizontal axis has a clamp with micrometric motion and sometimes a wheel used for turning the instrument.

Electric contact rings are placed at the ends of the horizontal axis. There should be enough contact rings to make possible the functioning of all the electric circuits going to the eyepiece when the instrument is adjusted for elevation. The wires from the contact rings to the eyepiece pass inside the telescope tube or underneath the protecting tube if there is any. The flanges are usually fastened symmetrically on both plates of the bearings to allow the electric circuits in both positions of the instrument to function.

14. Easing the Weight of the Horizontal Axis and Resetting the Instrument

The movable portion of the instrument, that is, the telescope and the horizontal axis, weighs 300-500 kilograms. Thus, there arises the problem of protecting the surfaces of the pivots against wear and tear and simplifying resetting of the instrument. The structural solution of one of these problems depends to a great measure on the solution of the other. It is important to keep the pivots in good shape because their iregularities affect the position of the line of sight of the instrument. Resetting the instrument is necessary to eliminate certain systematic instrumental errors.

To preserve the pivots from wearing, floating systems (reducing the pressure of the pivots on the bearings) are employed. Floating systems are divided into two types according to their construction (Fig. 18).

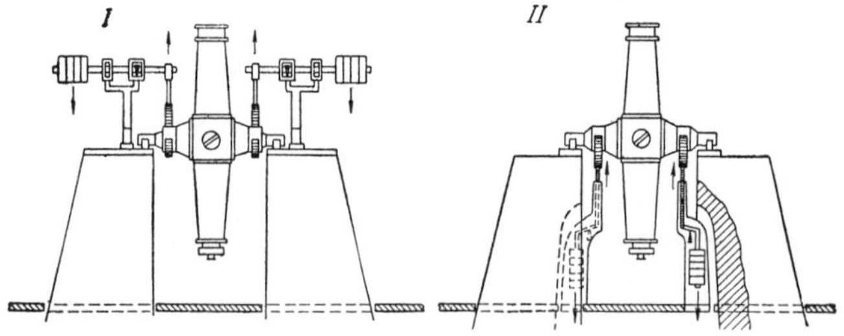

Fig. 18. Easing the weight of the horizontal axis.

The first type of floating is the simpler. At the top of each pillar of the instrument, there are supports for levers of the first kind. Weights are placed at the outer ends of the levers, and at the inner ends are hooks which grasp and support the horizontal axis. By regulating the weight and position of the weights on the levers, it is possible to have a significant part of the weight of the horizontal axis rest on the hooks and the pressure of the pivots on the bearings to be greatly diminished. Ordinarily, the pressure of the pivots on the bearings is fixed within three to five kilograms (sometimes up to ten kilograms) for each pivot. This pressure can be measured by lifting each end of the axis in turn over its bearing with a dynamometer. The pressure on the bearings must not be made too small because then the instrument will have no definite position on the bearings and will, as they say, "swim" on them. In order that the floating system may not influence the position of the horizontal axis of the instrument, which is determined by the position of the bearings, the weight-reducing hooks rest on hinges and hold the horizontal axis by means of friction rollers.

This is the most widely used system of floating; in it, resetting is carried out by means of a jack rolled along rails underneath the instrument. By means of a manual or mechanical crank, the upper part of the jack is lifted and two widely spaced forks hold the horizontal axis and lift it from the hooks of the floating system. In order to keep the hooks from following the axis because of the pressure of the weights, stopping devices are set up for the horizontal levers. After this, the instrument, now lying on the jack, is rolled along the meridian. Outside the pillars, it, along with the upper part of the jack, is rotated 180°; then it is rolled back and lowered onto the weight-reducing hooks and the bearings. After resetting the instrument, the micrometer clamp and the supports of the horizontal axis (which were unfastened before the shift) must be refastened. The jack is seen in Fig. 49.

Large transit instruments with straight telescope tubes can be reset without rolling the instrument out if the hooks are removed; but with meridian circles the graduated circles and the microscopes are obstacles to this type of resetting.

With a floating system of the second kind, the so-called columns are used. Two columns with forks and rollers at their upper ends support the horizontal axis of the instrument from below at two points. At their other ends, they stand on a circular platform on a level with the floor and isolated from it. The platform is the top of the resetting apparatus, which is set up on the base plate of the instrument. Weight-reducing levers are placed inside the columns together with counterweights or springs that act on these levers and reduce the pressure of the pivots on the bearings.

From a technical point of view, this construction is more complicated becaused the columns supporting the horizontal axis from below must not affect its position. Therefore, the columns must have the necessary degrees of freedom, which is considerably more complicated to attain than the hinged suspension of the hooks. On the other hand, resetting the instrument becomes simpler; there is no need of a separate jack. A lifting mechanism under the floor raises the instrument above the bearings by means of the columns, which are simultaneously extended; then the instrument is reset by rotation of the circular platform and is subsequently lowered onto the bearings.

With this method of resetting, the graduated circles can be placed closer to the tube. This makes it possible to reset the instrument between the pillars even if it has circles with relatively large diameters. It is important in principle to get the graduated circles closer to the line of sight because it reduces the errors that can arise when the line of sight and the graduated circle are displaced relative to each other as a result of the torsion of the horizontal axis. Ideally, the graduated circle should lie in the same plane as the line of sight.

Fig. 19. Meridian circle at time of resetting.

Floating and the first type of resetting were widely used in the past, for example, with the instruments Repsold worked with; among contemporary instruments, we may mention the seven-inch meridian circle at the U. S. Naval Observatory. The second type of construction is encountered more rarely; examples are the instruments at the Capetown, Uccle, and Moscow observatories. In the last case (Fig. 19), because of such a floating system, it is possible to reset the instrument automatically in twenty seconds, which makes it possible to observe stars from the pole to sixty degrees of declination with two positions of the instrument for one transit.

15. The Inclination of the Horizontal Axis and Its Determination

Ideally, the axis of rotation of the instrument should be horizontal and directed from east to west. The horizontal axis may deviate from the ideal position both in the vertical and in the horizontal planes. These deviations are errors made in setting up the instrument and are called the inclination and the azimuth of the horizontal axis.

Determination of the azimuth can be made only from observations of stars, about which we shall speak in Chapter IX. The inclination, on the other hand, is easily determined by laboratory methods with the help of a striding level.

The striding level consists of one or two glass tubes in casings placed on a special frame. The two legs of the striding level terminate in inverted V-bearings for setting the level on the pivots of the instrument. The graduated divisions on the glass tubes are usually two millimeters apart. In fundamental astrometry these divisions correspond to one, or occasionally to two seconds of arc; in the first case the radius of curvature of the glass tube is 400 meters; in the second case it is 200 meters. As a rule, levels are used which have a device to regulate the length of the bubble when this length changes with change in temperature.

The glass tube is enclosed in a hollow metallic cylinder with an aperture along the scale. The glass tube lies in it on two V-shaped posts against which it is pressed at the top by two springs. The metal tube is fastened in the outside casing of the level and it has reset shifts. Ordinarily, at one end of the metal tube there is a ball-and-socket joint and at the other a square shaft. In its side planes, there are four screws: two horizontal, for keeping the axis of the glass tube in the plane of the horizontal axis of the instrument, and two vertical, for correcting the inequalities of the legs of the striding level. Outside, the casing of the level is covered with a heat-insulating material so that large thermal deformations do not occur.

After the level is set on the pivots, it is necessary to check whether the axis of the level and the horizontal axis of the instrument are in the same plane. To do this, the frame with the level is rocked a little, without taking it from the axis, in the direction of the meridian. The immovability of the bubble guarantees the accuracy of the adjustment in this direction.

The striding level is hung on a special stand placed on the floor or on a pillar of the instrument (see frontispiece). A pulley or other such device makes it possible to lower the level smoothly onto the horizontal axis, to lift it and to reverse it on the axis. In order not to damage the pivots when the level is lowered onto them, its legs have rollers with springs on the ends. The rollers first come in contact with the horizontal axis of the instrument outside the working portions of the pivots. When the level is lowered further, the springs of the rollers are compressed, which lessens the pressure of the legs of the level on working portions of the pivots.

The reading of the positions of the ends of the bubble of the level on the scale is usually done by means of a small telescope. Recently, a photographic method of reading has become more common. This makes it possible to read the level accurately and objectively.

The principle of measuring the inclination is exceedingly simple. For this, readings are taken at the ends of the bubble in two positions of the striding level on the horizontal axis. If a_1, b_1 and a_2, b_2 are the readings at the ends of the bubble, the inclination is obtained from the formula

$$i = \frac{\alpha''}{4}[(a_2 + b_2) - (a_1 + b_1)],$$

where α'' is the value of a division of the level.

We need only to decide in what order we should take these readings so that the sign of i will satisfy the following condition: $i > 0$ if the west end of the horizontal axis is the higher.

Similarly, the unequal legs of the striding level distort the readings of the bubble ends, but this is eliminated in the difference given in the above formula. However, it is convenient if the inequality of legs is not too large. The determination of the inclination is carried out always at the same altitude of the instrument so that the effect of pivot irregularities on the obtained value of inclination will always be the same.

In addition to the striding level, a hanging level similar to the type in small transit instruments with a broken-telescope tube is used. With this level, it is possible to determine the inclination when the telescope tube is vertical, which permits obtaining the collimation from readings of the mercury surface at the nadir.

The level seems to be the most unreliable and capricious part of every astrometric instrument. Since attempts to replace it by other inclination measuring devices, in particular by elastic quartz threads, were not successful, one is obliged to investigate it in detail and study the behavior of the level under various conditions.

16. The Study of the Level

The investigation of the level is made with a device called a "level-trier." (For a detailed description, we refer the reader to any course in practical astronomy.) The level is set on the trier so that its axis is perpendicular to the trier's axis of rotation. The testing process consists in inclining the trier together with the level to an angle by turning the trier micrometer screw. Turning the screw of the trier successively over a constant number of divisions drives the bubble along the whole scale from one end to the other. At each setting, after the bubble of the level has settled down, which usually happens after one to two minutes, the positions of the ends of the bubble a_i and b_i are read from the scale and the position p_i of the trier screw is noted. Altogether, about ten to twelve points are measured along the scale of the level. In order to eliminate the linear component of the variation of the setting with time, the succession of measurements is repeated in the opposite direction using the same readings of the screw.

The basic equation relating the screw readings to the value of one level division may be written in the following manner:

$$p_i = p_0 + \alpha x_i,$$

where α is the actual value of the level division, x_i is the reading of the midpoint of the bubble; that is, half the sum of its end readings, and p_0 is the reading of the trier screw corresponding to $x_i = 0$. Solving this system of equations directly by means of the method of least squares, it is possible to determine α and p_0. It is better to simplify the equations by eliminating the unnecessary and unknown quantity p_0 according to the method of Cauchy. For this purpose, one forms the mean equation $\bar{p} = p_0 + \alpha\bar{x}$ and substracts it from each equation of the system. Then p_0 is eliminated and the equations for determination of α take on a simpler form:

$$p_i - \bar{p} = \alpha(x_i - \bar{x}), \tag{4}$$

The value of the divisions of the level are then determined according to the formula

$$\alpha = \frac{\sum (p_i - \bar{p})(x_i - \bar{x})}{\sum (x_i - \bar{x})^2}.$$

However, it is easier to apply the analysis due to Wanach (Section 17).

Usually, in order to simplify the analysis, α is determined in the number of divisions of the trier screw which is subsequently multiplied by the value of the screw division. It is useful, to begin with, to construct a graph relating x_i and p_i. It is easy to see that α is the mean value of the level division, since it is equal to the slope of the line drawn in the best way through the plotted points. An analysis of the deviation of points from this line may give a notion of the reliability of the glass tube and of the experimental precision.

The average value of the division of the level depends upon the temperature and upon the length of the bubble. Even though, in practice, the latter can be regulated, in view of the fact that the length varies directly with a change in temperature, it is not possible to keep it constant. During a change in temperature of 1°, the length of the bubble in the seconds level changes by 1.0–1.5 mm. In this way, the temperature influences the value of one division of the level while it deforms the glass tube with its brass container and thus indirectly changes the length of the bubble. The dependence of the average value of the division upon the length of the bubble can be explained by the irregularities of the internal surface of the glass tube. Moreover, these irregularities change the value of the divisions for different sections of the scale for the same bubble length. The smallest effect of the glass tube, called "aging," is a variation in the value of the divisions of the level with time.

To determine the dependence of the mean value of the divisions of the level on temperature, the trier is placed either in a thermostatically controlled chamber, or in the pavilion under ordinary open-air conditions. Dependence on temperature is usually linear, and only very rarely is it necessary to use the quadratic term

$$\alpha_i = \alpha_0 + k_1 t_i + k_2 t_i^2.$$

Here t_i is the temperature, and k_1 and k_2 are constant coefficients.

By solving the equations corresponding to different temperatures t_i by the method of least squares, it is possible to obtain the values of the coefficients k_1 and k_2. For a more certain determination of the coefficients, it is necessary to investigate the level within a larger temperature range, for example, from +25° to −25°C. The quantity k_1 for a seconds level has a range of 0″.001–0″.002 per degree. The determination of the temperature dependence must be carried out at constant bubble length.

Astronomers always try to keep the length of the bubble constant, ordinarily at about two-thirds the length of the level scale (the longer the bubble, the more sensitive it is). However, the difficulty of adjusting the length of the bubble, which is known to vary somewhat with temperature, forces us to investigate the dependence of α upon the length of the bubble l. The dependence of α upon l is ordinarily linear, analogous to the dependence upon temperature. The quadratic term is very rarely taken into account.

Levels showing greater dependence of the average value of the division on the bubble length also have varying values of divisions along the scale. Such levels should be replaced in view of the complications and difficulties in taking account of this phenomenon.

Now it is appropriate to describe briefly the possibilities of approximate investigation of the irregularities of the inner surface of the level glass tube. We give here, as an example, the effects of these irregularities as estimated according to the method of the Pulkovo astronomer A. S. Vasil'yev.

The investigation of the level according to this method is usually carried out by a forward and backward movement of the bubble. The readings are produced in equal time intervals Δt and the identical screw readings of the trier with the bubble proceeding in either direction.

Equations relating the readings of the trier to the readings of the midpoint of the bubble during its forward motion may be written in the following form

$$\frac{1}{\alpha} p_i = \frac{1}{\alpha} p_0 + x_i + \Delta t (i - 1) z + p \qquad (i = 1, 2, \ldots, n),$$

while for the backward motion we have

$$\frac{1}{\alpha} p_i' = \frac{1}{\alpha} p_0 + x_i' + \Delta t (2n - 1 - i) z - p \quad (i = 1, 2, \ldots, n),$$

where z includes the linear variation of position with time, and p is the effect of the imperfections of the internal surface of the glass tube at the bubble midpoint. It is assumed that the effect of the irregularities of the glass tube on the bubble shift has an opposite sign depending on the direction of bubble motion. Inasmuch as the average effect of the irregularities of the glass tube is estimated, the quantity p is considered to be constant.

If the differences of the two equations are formed in pairs for the same screw position of the trier, we obtain the following system of equations for the determination of p and z:

$$2p + 2\Delta t (i - n) z = x_i' - x_i.$$

The quantity p for good levels should not exceed 0.05 of a scale division of the glass tube. The quantity z ordinarily does not exceed 0.03 of a division.

Let us substitute the values p and z obtained by the method of least squares in the conditional equations. Ignoring accidental errors, the resulting disagreements may be considered as errors in the position of the bubble midpoint. The latter may be regarded as the result of the composite influence of the glass tube irregularities which occur at the ends of the bubble. With a different bubble length, the same reading of its midpoint will be distorted differently. Due to these circumstances, the value of the division of the level will depend as much upon the length of the bubble as upon the part of the glass tube used.

Up to the present, there are no rational methods of determining and accounting for the irregularities of the internal surface of the glass tube.

17. The Wanach Method for Determining the Mean Value of the Level Division

A large saving in time can be achieved by solving equations (4) according to the method of Wanach. It can be easily applied whenever it is necessary to determine only the slope of the straight line which best fits the observed points. The value of α obtained by this method corresponds completely to the solution by the method of least squares.

Let us transform equations (4) into the form

$$x_i - \bar{x} = \frac{1}{\alpha}(p_i - \bar{p}), \qquad i = 1, 2, 3, \ldots, n,$$

then, solving according to the method of least squares, we obtain

$$\frac{1}{\alpha} = \frac{\sum (x_i - \bar{x})(p_i - \bar{p})}{\sum (p_i - \bar{p})^2}.$$

The resetting of the head of the trier screw is always made by a constant number of divisions. In order to simplify the derivation, let us assume that one turn of the screw corresponds to a shift of one division; i.e., $\Delta p = 1$. (Further on, we shall show how the final formulas change in the case when $\Delta p = k$.) The quantities $(p_i - \bar{p})$ are then represented then by a series of $\ldots -2.5, -1.5, +0.5, +1.5, +2.5, \ldots$ if the number of positions are even ($n = 2m$), or $-3, -2, -1, 0, +1, +2, +3, \ldots$ if the number is odd ($n = 2m + 1$). In both cases, the following equality is justified:

$$p_i - \bar{p} = -(p_{n-i+1} - \bar{p}).$$

Using this relationship, the numerator may be changed in the following manner:

$$\sum_{i=1}^{n} (x_i - \bar{x})(p_i - \bar{p}) = (x_1 - \bar{x})(p_1 - \bar{p}) + \cdots$$
$$\cdots + (x_i - \bar{x})(p_i - \bar{p}) + \cdots + (x_{n-i+1} - x)(p_{n-i+1} - \bar{p}) + \cdots$$
$$\cdots + (x_n - \bar{x})(p_n - \bar{p}) = (x_1 - x_n)(p_1 - \bar{p}) + \cdots$$
$$\cdots + (x_i - x_{n-i+1})(p_i - \bar{p}) + \cdots + (x_n - x_{n-m+1})(p_m - \bar{p}) =$$
$$= \sum_{i=1}^{m} (x_i - x_{n-i+1})(p_i - \bar{p})$$

and the denominator:

$$\sum_{i=1}^{n} (p_i - \bar{p})^2 = 2\sum_{i=1}^{m} (p_i - \bar{p})^2 = 2\left[\left(\frac{1-n}{2}\right)^2 + \left(\frac{3-n}{2}\right)^2 + \left(\frac{5-n}{2}\right)^2 + \cdots + (-1)^2\right]$$

Since the final formulas are somewhat different for the cases $n = 2m$ and $n = 2m + 1$, we consider these two cases separately.

For the first case the numerator will be equal to

$$\frac{1}{2}\sum_{i=1}^{m}(x_i - x_{n-i+1})(2i - 1 - 2m),$$

and the denominator can be transformed as follows:

$$\sum_{i=1}^{n}(p_i - \bar{p})^2 = 2\left[\left(\frac{1}{2}\right)^2 + \left(\frac{3}{2}\right)^2 + \ldots + \left(\frac{2m-1}{2}\right)^2\right] = \frac{2}{4}[1^2 + 3^2 + \ldots +$$

$$+ (2m-1)^2] = \frac{1}{2}\cdot\frac{1}{3}m(4m^2 - 1) = \frac{1}{2}\left[(n-1)^2 + (n-3)^2 + (n-5)^2 +\right.$$

$$\left. + \ldots + 5^2 + 3^2 + 1^2\right] = \frac{1}{2}\cdot\frac{(n-1)n(n+1)}{6} = \frac{n(n^2-1)}{12}.$$

For the second case the numerator will be equal to

$$\sum_{i=1}^{m}(x_i - x_{n-i+1})(i - m - 1),$$

and the denominator

$$\sum_{i=1}^{m}(p_i - \bar{p})^2 = 2[1^2 + 2^2 + \ldots + (m-1)^2] = 2\frac{(m-1)m(2m-1)}{1\cdot 2\cdot 3} =$$

$$= \frac{1}{2}\left[(n+1)^2 + (n-3)^2 + (n-5)^2 + \ldots + 6^2 + 4^2 + 2^2\right] = \frac{n(n^2-1)}{12}$$

In this way, it is finally possible to represent, respectively, the value of a division of the level for the cases $n = 2m$ and $n = 2m + 1$ by

$$\left.\begin{aligned} \alpha &= \frac{n(n^2-1)}{6}\,\frac{1}{\sum\limits_{i=1}^{m}(x_i - x_{n-i+1})\,(2i - 1 - 2m)}; \\ \alpha &= \frac{n(n^2-1)}{12}\,\frac{1}{\sum\limits_{i=1}^{m}(x_i - x_{n-i+1})\,(i - m - 1m)}. \end{aligned}\right\} \quad (5)$$

The quantities $\sum\limits_{i=1}(x_i - x_{n-i+1})(2i - 1 - 2m)$ for an even number; and $\sum\limits_{i=1}^{m}(x_i - x_{n-i+1})(i - m - 1)$ for an odd number of settings are computed very easily, because they represent, by themselves. the sum of the products of the differences of the quantities x_i. They are symmetrically distributed around the mean setting and multiplied in each case by their corresponding number of settings, that is, by 1, 3, 5, ... or 1, 2, 3 ....

If the setting is not made for every division of the trier but for k divisions, then it is necessary to replace everywhere $(p_i - \bar{p})$ by $(p_i - \bar{p})k$; that is, coefficients in (5) will take the form

$$K = \frac{n(n^2-1)}{6} \text{ for the case } n = 2m$$

and

$$K = \frac{n(n^2-1)}{12} \text{ for the case } n = 2m+1.$$

These coefficients may be computed once, and represented in a two-entry table with arguments n and k. However, in practice, it is more convenient to construct a table with the argument $4K$ multiplied in addition by the value of one division of the trier screw. The convenience of using the argument $4K$ arises from the analysis of the measurements as recorded in the journal, provided one does not use the midpoint of the bubble x_i but the sum of the end readings $(a_i + b_i)$; this sum is added to the same quantity obtained from the reverse direction of measurement.

18. Irregularities of the Pivots

The pivots of the meridian instrument must be round, coaxial cylinders with equal diameters. The horizontal axis of the instrument is defined in this ideal case as a straight line joining the centers of the working cross section of the pivots. During rotation of an ideal instrument, the horizontal axis will keep a constant position in space, and, consequently, the projection of the line of sight will describe a large circle on the celestial sphere.

Irregularities of the pivots shift the horizontal axis of the instrument, changing its inclination and azimuth. Consequently, due to the irregularities of the pivots the line of sight will not describe a geometrical circle on the celestial sphere, but some curve, close to a circle, and the observed transit time of a star will not correspond to the true transit time; therefore, a study of the irregularities of the pivots is necessary for observations of the right ascensions. It is possible to visualize irregular pivots which cause only continuously progressive displacements of the horizontal axis. But, in practice, this is not very probable.

Irregularities of the pivots do not influence the reading of the circle, since the displacement of the divided circle with respect to the reading microscopes caused by these irregularities may be eliminated by using the diametrically opposite microscopes. In principle, insofar as the irregularities of the pivots change i and k, they influence the observed zenith distance z in the second-order terms only. However, in view of the small size of the irregularities of the pivots, this effect is not perceptible.

The production of straight pivots is technically difficult, in spite of the efforts to make them as precise as possible. The deviation of the shape of the pivot from a circle does not exceed one micron in modern instruments; and in some cases not even 0.1 micron. But even these small irregularities cause noticeable displacements in the line of sight. Thus, with an axis of 100 cm and an irregularity of the pivot of 1 micron, this displacement is as much as $0''.014$.

The elimination of errors caused by the irregularities of pivots appears to be one of the important questions in the determination of right ascensions.

Irregularities of the pivots require greater precision in the definition of the motion of the horizontal axis, as long as concepts of the center and radius of the pivots remain undefined. During the rotation of an irregular pivot, no point in its operating cross section is considered immobile. Consequently, one introduces the notion of an "ideal pivot" in relation to which one determines the corrections for an irregular pivot. Figure 20 shows a contour of the working cross section of the pivot, the irregularities of which are enlarged many times for clarity. The "ideal pivot" is represented by a circle. Let ρ be the radius-vector of a point on the contour of the pivot, R the radius of the "ideal pivot," and O its center. The quantities $\Delta\rho = \rho - R$ are the essential part of the irregularity of the pivots which displaces the horizontal axis during the rotation of the instrument and brings about the variations in its inclination and azimuth.

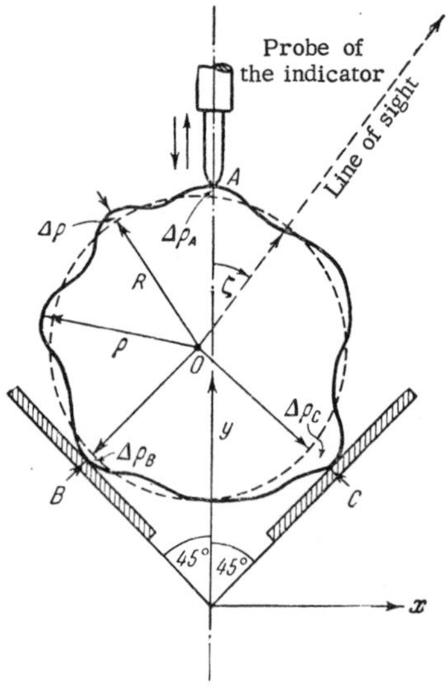

Fig. 20. Irregularities of a pivot.

The quantity ζ for the correction of the irregularity of pivots is introduced in the following way. Let us extend the radius-vector from the assumed center of the pivot to its upper point while the instrument is pointed towards the zenith. We shall read off the quantity ζ, the angle of rotation of this radius-vector relative to the vertical line. The quantity ζ is counted from 0° to 360° in the direction from the zenith to the south.

The irregularities $\Delta\rho$ of the figure of the pivot at points adjacent to the bearing bring about a displacement of the center of the ideal pivot. This displacement may be divided into its components Δx_ζ and Δy_ζ along the horizontal and vertical axes, respectively. These are then the variations of the azimuth and the inclination of the horizontal axis caused by the irregularities of one pivot. The problem of investigation of pivots is reduced to the determination of the quantities Δx_ζ and Δy_ζ (for each of them) during different elevations of the instrument. Some methods of investigating immediately give the total effect of the irregularities of both pivots on the azimuth (Δk_ζ) and the inclination (Δi_ζ) of the horizontal axis.

The system of corrections Δx_ζ and Δy_ζ depends upon the assumed position of the center of the ideal pivot and its radius. Mathematically, it is possible to formulate conditions determining the ideal pivot in different ways. It is worthwhile to design the ideal pivot so that the corrections Δx_ζ and Δy_ζ are least.

Insofar as it is necessary to determine the position of the center and the radius of the ideal pivot, it is also necessary for the determination of the system of corrections to impose two conditions on the quantities $\Delta\rho$. One can assume, for example, the following conditions: $\sum \Delta\rho_\zeta = 0$ and $\sum \Delta\rho_\zeta^2 \rightarrow$ min. In this case, we take as the center of the ideal pivot the center of inertia of the points on the curve of the working cross section of the pivot, for which $\sum \rho_\zeta^2 \rightarrow$ min. Actually, we expand the sum $\sum \rho_\zeta^2$ by substituting $R + \Delta\rho_\zeta$ for ρ_ζ since $\sum R^2 =$ const, and, according to the conditions $\sum \Delta\rho_\zeta = 0$ and $\sum \Delta\rho_\zeta^2 \rightarrow$ min, $\sum \rho_\zeta^2$ is also a minimum:

$$\sum \rho_\zeta^2 = \sum (R + \Delta\rho_\zeta)^2 = \sum R^2 + 2R \sum \Delta\rho_\zeta + \sum \Delta\rho_\zeta^2.$$

Again, the bearings of the instrument, like the pivots, are not geometrically ideal either. Since they are made from a softer material than the pivots, they wear off and touch the pivots not at points but at surfaces; generally speaking, this effect varies from one small surface to another. Hence, for complete understanding of the influence of the irregularities of the pivots, it is necessary to consider them together with the bearings. One realizes that the displacement of the horizontal axis, and consequently of the line of sight, comes about through the composite action of pivot and bearing irregularities. The interaction of the pivot with different bearings, after resetting the instrument, may be different, and therefore may in each case cause different errors in observation. For this same reason, one analyzes every pivot on every bearing; that is, one investigates the pivots in both positions of the instrument.

An important, although rather refined question, comes up: Do irregularities of pivots similarly show up while in contact with the striding level and with the bearings; that is, is it possible to use the same corrections obtained from the level readings, for correcting the inclination and for correcting the terms in Mayer's formula? This may be significant if one desires to eliminate completely the observational systematic errors.

Before proceeding to the description of methods for determining the irregularities of pivots, let us consider the inequality and the departure from coaxiality of the pivots.

Inequality of the pivots causes distortion of the inclination determined by a striding level. This distortion has different signs in the two positions of the instrument. Thus, the difference of the inclinations for the two positions of the instrument (circle-east, circle-west) will be equal to twice the effect of the inequality of the pivots $\Delta i'$:

$$i_E - i_W = 2\,\Delta i' = 2\,\frac{D_1 - D_2}{l\sqrt{2}},$$

where l is the length of the horizontal axis between the working cross sections, D_1 and D_2 are pivot diameters. After the determination of the quantity $\Delta i'$, it is necessary to apply this correction to the inclination determined by the striding level.

To avoid the effect of the irregularities of the pivots these investigations, as well as the determinations of the inclination, should be carried out so that the pivots of the instrument touch the bearings on the same points in both positions of the instrument; otherwise, one must introduce the corrections for the irregularities of the pivots, Δi_ζ, which reduce the readings of the striding level to the readings of the ideal pivots.

The axes of cylindrical pivots must be in the same straight line. Departure from coaxiality of the pivots will cause their working cross sections—i.e., the totality of points in contact with the bearings—to be ellipses rather than cylindrical and out of the perpendicular to the pivot axes. This will cause regular beats of the horizontal axis.

Departure from coaxility is investigated with a small striding level, in which the distance between its legs does not exceed the length of the pivot. By placing it on each pivot, its inclination is determined at different elevation positions of the instrument. The sinusoidal variation of the inclination of each pivot, and the coincidence of phase and amplitude of the total inclination variation over both pivots with the general character of the variation of the inclination of the horizontal axis, show the departure from coaxiality. During these investigations, it is assumed that the cross sections of the pivots over which the legs of the level are placed have a shape similar to the working cross sections. This assumption is, generally speaking, approximately true, as long as the pivots are finished with a wide polishing ring.

For large departures from coaxiality, as in the case of large irregularities, the pivots must be repolished.

During investigation of pivots, the following can be determined:

1) The shape of each pivot with the subsequent derivation of corrections for its irregularities.

2) The displacement of the center of each pivot Δx_ζ and Δy_ζ.

3) The displacement of the horizontal axis Δi_ζ and Δk_ζ.

19. Investigation of the Shape of Pivots

The first group of methods studies the shape of pivots. In these methods, one usually measures the vertical displacement of the upper point of the pivot. For these measurements, indicators are used to record linear displacements and magnify small shifts of the measuring probe by mechanical or optical means. The precision of the indicators has already been found inadequate for modern quality pivots; therefore, it is better to use interferometers which have the accuracy of ±0.01 micron for one reading. The indicator is fixed on a special stand above the pivot, so that the measuring probe moves in the vertical direction and touches an upper point of the pivot. During the rotation of the instrument, the irregularities of the pivot will displace the measuring probe up and down and the reading of the indicator will vary accordingly.

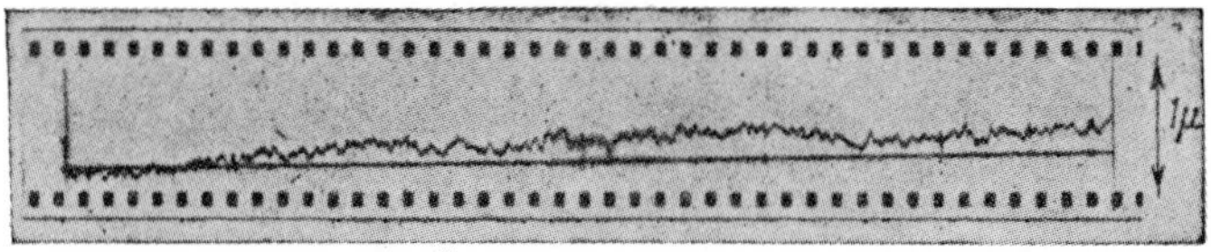

Fig. 21. Photographic recording of the irregularities of a pivot.

Figure 21 shows a photographic record (enlarged 20,000 times) of the displacements of the measuring probe during a complete revolution of the pivot. The record is obtained by an indicator, which operates on the principle of optical multiplication (see schematic diagram in Fig. 22). The displacements of the measuring probe A change the inclination of the mirror B. Before the ray of light reaches the film, it undergoes multiple reflections between the mirrors B and C, and leaves this path at a much larger angle than the angle of inclination of mirror B.

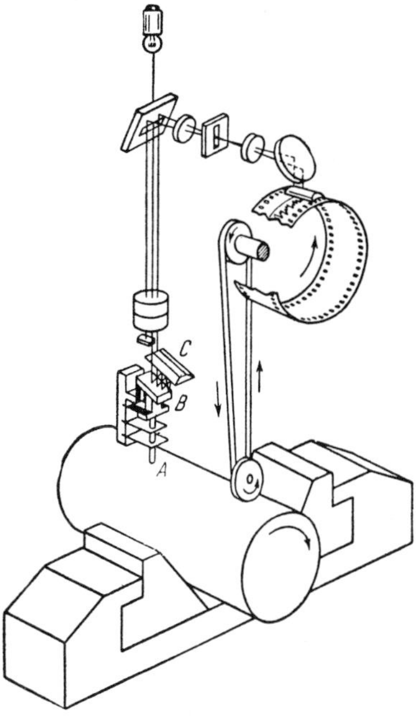

Fig. 22. The scheme of an optical indicator.

Visual measurements of the probe displacements are obtained as the instrument is turned step by step around the horizontal axis by an angle of $\frac{360°}{n}$ until a full revolution of 360° is completed. The deviation of each indicator reading A_ζ from the average value A_0 for the whole cycle is the consequence of the pivot irregularities at the three points of contact: two of these are between the pivot and the bearing surfaces (see Fig. 20), and the third is between the pivot and the probe of the indicator:

$$A_\zeta - A_0 = \Delta\rho_A + a\,\Delta\rho_B + b\,\Delta\rho_C.$$

The coefficients a and b depend upon the angle between the bearing surfaces, also upon whether the measuring probe is placed in the bisectrix of this angle. The latter condition is usually fulfilled, and if the pivot irregularities are small and the angle between the surfaces is nearly 90°, the coefficients may be assumed to be equal to sin 45° $= \frac{1}{\sqrt{2}}$.

The quantities $\Delta\rho_\zeta$ may be determined easily by setting the instrument through 45°, as was suggested by Moreau and Verbaandert. At other angles, the irregularities of the pivots should be represented by a Fourier series with determined coefficients. The measurements obtained by setting the instrument for every 45° give eight equations of the form

$$\Delta A_\zeta = \Delta\rho_\zeta + \frac{1}{\sqrt{2}}\left(\Delta\rho_{\zeta+135^\circ} + \Delta\rho_{\zeta-135^\circ}\right);\ \zeta = 0^\circ, 45^\circ, \ldots, 315^\circ. \tag{6}$$

It is easy to show that, of the eight equations, only six are independent. The indeterminacy of this system arises from the fact that the quantities $\Delta\rho_\zeta$ depend upon the position of the center and the size of the radius of the ideal pivot. In order to avoid the indeterminacy, it is necessary to impose some conditions.

For these conditions, one may assume $\sum \Delta\rho_\zeta = 0$ to define the radius, and $\sum \Delta\rho_\zeta^2 \to \min$ to define the position of the center of an ideal pivot. The first condition is already fulfilled, because we have taken A_0 as the average reading. Actually, if we sum up all the equations (6), we obtain $\sum \Delta A_\zeta = \left(1 + \frac{2}{\sqrt{2}}\right) \sum \Delta\rho_\zeta$; inasmuch as $\sum \Delta A_\zeta$ is equal to zero, as the sum of the deviations from the mean, $\sum \Delta\rho_\zeta$ is also equal to zero.

The second condition must be fulfilled in the solution of the system of equations.

The solution of equations (6) may be carried through by the method of correlates, which makes it possible to solve a system of linear equations with an excess of unknowns if the condition of the minimum sum of squares is observed. The idea of the method of correlates consists in the following: Given a system of p linear equations with n unknowns, where $p < n$:

$$\begin{aligned} a_1x_1 + a_2x_2 + \ldots + a_nx_n &= l_1, \\ b_1x_1 + b_2x_2 + \ldots + b_nx_n &= l_2, \\ \cdots\cdots\cdots\cdots\cdots & \\ p_1x_1 + p_2x_2 + \ldots + p_nx_n &= l_p, \end{aligned}$$

it is necessary to find x_i so that $\sum x_i^2 \to \min$. We shall look for a solution of the form

$$\begin{aligned} x_1 &= a_1K_1 + b_1K_2 + \ldots + p_1K_p, \\ x_2 &= a_2K_1 + b_2K_2 + \ldots + p_2K_p, \\ &\cdots\cdots\cdots\cdots\cdots \\ x_n &= a_nK_1 + b_nK_2 + \ldots + p_nK_p, \end{aligned}$$

where K_i are the unknown coefficients. Let us write down the normal equations for this system of n linear equations

$$[aa]K_1+[ab]K_2+\dots+[ap]K_p=l_1,$$
$$[ba]K_1+[bb]K_2+\dots+[bp]K_p=l_2,$$
$$\dots\dots\dots\dots\dots\dots\dots$$
$$[pa]K_1+[pb]K_2+\dots+[pp]K_p=l_p.$$

By solving this system, we can determine the values K_i and, through them, x_i.

The figure of the pivot may be represented in the form of a Fourier series

$$\Delta\rho_\zeta=\Delta\rho_0+b\sin(2\zeta+B)+c\sin(3\zeta+C)+\dots$$

There is no first-order term under the assumed conditions. As Whink showed, to assure the determination of the coefficients of this formula, it is necessary to make one more sequence of measurements with the indicator probe placed along the pivot radius and forming an angle γ with the vertical direction. The weights of the unknowns $b, c, \dots B, C, \dots$ are not found to be equal, but change with the variations in the angle γ.

Having determined the irregularities of the shaft $\Delta\rho_\zeta$ by this or any other method, it is possible to find the displacements Δx_ζ and Δy_ζ of the center of the shaft according to the formulas

$$\Delta x_\zeta=\frac{1}{\sqrt{2}}(\Delta\rho_{\zeta+135^\circ}-\Delta\rho_{\zeta-135^\circ});\quad \Delta y_\zeta=\frac{1}{\sqrt{2}}(\Delta\rho_{\zeta+135^\circ}+\Delta\rho_{\zeta-135^\circ}).$$

The angle between each of the bearing surfaces and the vertical line can be assumed to be equal to 45°.

By finding the analogous quantities for the second pivot, and by forming the differences of displacements of the centers of both pivots for equal values of the argument ζ, we obtain the displacements Δi_ζ and Δk_ζ of the horizontal axis. By expressing them in seconds of time, and by multiplying them with the corresponding coefficients I and K of the Mayer formula, the corrections of the observed moments for the pivot irregularities may be obtained.

The results of the investigation by these methods cannot be considered completely adequate for determining the corrections for the pivot irregularities. In these methods, the interpretations of the obtained quantities are oversimplified, since it is assumed that the irregularities show up in the same way during contacts of the bearings and the measuring probe with each other. However, these methods are satisfactory for a preliminary study of the quality of the pivots; for example, in the factory examination of pivots during the polishing process.

Methods for direct determination of Δx_ζ and Δy_ζ are more rigorous. With the method described in Section 21, the difference between the interaction of the pivot with the indicator probe and the bearings may be determined.

20. Methods of Determining Pivot Center Displacement Due to Irregularities

The classical method in this group is called the method of Challis. A very small round mark is placed inside the pivot in the plane of its working cross section and as near as possible to the center. One of the ways to do this is to precipitate on a clean glass surface some mercury vapors which will form microscopic regular balls. The placing of the mark in the plane of the working cross section is necessary so that the irregularities of the second pivot will not affect its displacements.

In front of the mark, on a bracket or stand attached to the pillar of the instrument, is placed a measuring microscope whose axis is directed along the horizontal axis of the instrument. The microscope is equipped with a double micrometer, the screws of which are oriented along the vertical and horizontal directions.

If the pivot were ideal, then during the rotation of the instrument the mark M would describe a circle in the line of sight of the microscope, with a radius equal to the distance OM between the center O and the mark. In the presence of irregularities, this circle is transformed into some irregular closed curve. Considering this curve to consist of the circle defined by the distance from the measuring mark to the center of the ideal pivot and of the displacements of the horizontal axis resulting from irregularities of the pivots, the center of the ideal pivot is assumed to be the point which has no circular motion.

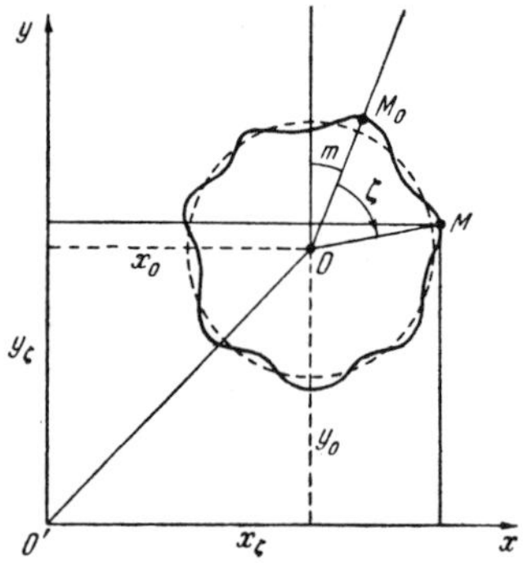

Fig. 23. Shifts of the mark in the investigation by the Challis method.

The instrument is turned by a constant angle $\frac{360°}{n}$, and at each step a correction is made by means of the crossed threads in the twin microscope on the mark. Thus, two sets of quantities x_ζ and y_ζ are obtained which are the readings of the horizontal and vertical micrometer screws. We read off the quantities x_ζ and y_ζ to the nearest fraction of a millimeter; the values of the micrometer revolutions are determined by measuring some scale with the same microscope without changing its focal length. Figure 23 shows the center of the ideal pivot O, referred to the measuring system of coordinates, and also the trajectory of the mark as well as the circle which would be described by the mark for a true pivot. Let M_0 be the position of the mark when $\zeta = 0$, let M be an arbitrary position. The angle subtended by the radius-vector OM_0 with the vertical, and counted in the same direction as ζ, is designated by m. Then, the measured coordinates of the mark may be represented by the following formulas:

$$x_\zeta = x_0 + r\sin(m+\zeta) + \Delta x_\zeta;\quad y_\zeta = y_0 + r\cos(m+\zeta) + \Delta y_\zeta,$$

where x_0 and y_0 are the coordinates of the center of the ideal pivot and r is the distance between the mark and the center of the ideal pivot.

From these measurements we obtain a system of $2n$ equations with $2n+4$ unknowns (Δx_ζ, Δy_ζ, r, m, x_0, y_0). For the solution it is necessary to impose conditions on the unknowns. If it is assumed that $\sum \Delta x_\zeta = 0$ and $\sum \Delta y_\zeta = 0$, then it is possible to determine $x_0 = \bar{x}_\zeta$ and $y_0 = \bar{y}_\zeta$ upon adding up separately the equations for Δx_ζ and for Δy_ζ and by assuming an even distribution of the argument $(m+\zeta)$ along the circle.

After the elimination of the unknowns x_0 and y_0, the equations for the determination of Δx_ζ and Δy_ζ take the form

$$\left.\begin{aligned} x_\zeta - \bar{x}_\zeta &= r\sin(m+\zeta) + \Delta x_\zeta, \\ y_\zeta - \bar{y}_\zeta &= r\cos(m+\zeta) + \Delta y_\zeta. \end{aligned}\right\} \tag{7}$$

For the determination of the unknowns r and m we transform these equations

$$x_\zeta - \bar{x}_\zeta = r\sin m\cos\zeta + r\cos m\sin\zeta + \Delta x_\zeta,$$
$$y_\zeta - \bar{y}_\zeta = r\cos m\cos\zeta - r\sin m\sin\zeta + \Delta y_\zeta.$$

Let us introduce new unknowns $r\sin m = s$ and $r\cos m = t$, and write down for their determination the normal equations

$$\sum (x_\zeta - \bar{x}_\zeta)\cos\zeta = s\sum \cos^2\zeta + t\sum \sin\zeta\cos\zeta + \sum \Delta x_\zeta\cos\zeta,$$
$$\sum (y_\zeta - \bar{y}_\zeta)\sin\zeta = s\sum \sin\zeta\cos\zeta + t\sum \sin^2\zeta + \sum \Delta y_\zeta\sin\zeta.$$

Let us impose two more conditions:

$$\sum \Delta x_\zeta\cos\zeta = 0 \text{ and } \sum \Delta y_\zeta\sin\zeta = 0.$$

Then it is possible to determine the unknowns s and t:

$$s = \frac{2}{n}\sum (x_\zeta - \bar{x}_\zeta)\cos\zeta,\quad t = \frac{2}{n}\sum (y_\zeta - \bar{y}_\zeta)\sin\zeta,$$

and through them, the quantities r and m. We are interested in obtaining individual corrections; they will appear as residuals of the conditional equation (7). Subsequently we obtain the differences of the corresponding quantities Δy_ζ and Δx_ζ for the two pivots and thus determine the variations Δi_ζ and Δk_ζ of the inclination and the azimuth of the horizontal axis which result from irregularities of the pivots.

During one cycle of measurements by the Challis method the measuring microscope must remain stationary. The duration of one cycle may be shortened if, by choosing an appropriate optical system, the displacements of the mark are photographed. The mark should not be placed too near the center of the pivot, otherwise the track of the mark on the photographic plate will be distorted by photographic effects. The rectangular coordinates of the points on this curve corresponding to the uniformly increasing angle ζ are measured with an ordinary measuring device. A continuous record of the effects of the irregularities of the pivots may reveal the presence of displacements of the center of the pivot in the form of narrow maxima and minima. The latter may be omitted when using a visual method of determination at discrete instrumental positions which differ from each other by a fairly large constant angle. It is more convenient to measure not the points on the continuous curve, but a series of separate points which may be quickly obtained by a photographic method, even with small angles of rotation of the instrument.

21. Determination of the Displacement of the Center of the Pivot by Means of an Indicator

We assume that the indicator probe, measuring only the vertical displacements, touches the surface of the pivot with a plane inclined to the axis of the probe by some angle $90°-\gamma$ [Fig. 24, a] ($2\gamma = \frac{360°}{n}$, where n is a predetermined number of positions during one series of measurements; for example: $\gamma = 6$, 7.5 or 9°). The point of contact will then subtend an angle γ with the highest point of the pivot.

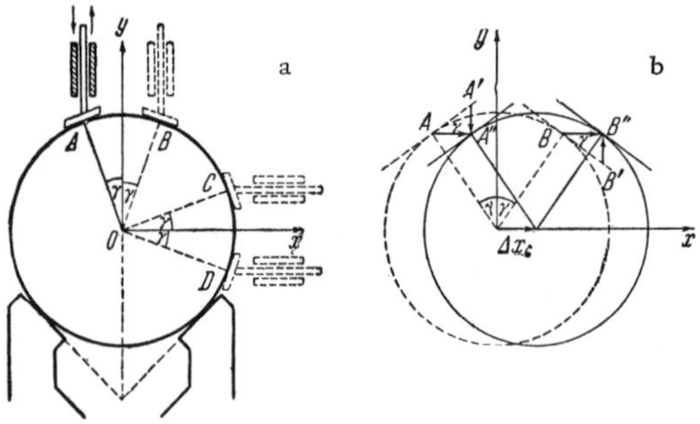

Fig. 24. The investigation of the shifts of a pivot with an indicator.

The indicator readings will contain the total displacement of the pivot center Δy along the vertical direction. The horizontal displacement Δx also changes the indicator reading, when compared with the ideal pivot reading, by an amount $\Delta x \tan\gamma$. This is easy to see from Fig. 24, b, where the dotted line shows the original pivot position and the solid line the displacement due to the quantity Δx. To avoid ambiguity, we assume Δx increasing to the right and Δy increasing upwards. The displacement of the plane of contact gives the vertical component (see Fig. 24, b):

$$A'A'' = \Delta x_\zeta \tan\gamma$$

(from the triangle $A'A''A$, where $AA'' = \Delta x_\zeta$ and angle $A'AA'' = \gamma$).

Finally, the reading along the indicator scale at any pivot position may be rewritten in the form

$$A_\zeta = A_0 + \Delta y_\zeta + \Delta x_\zeta \tan\gamma + \Delta\rho_\zeta \sec\gamma, \tag{8}$$

where the effect on the indicator reading of the irregularity of the pivot at the point in direct contact with the probe is $\Delta\rho_\zeta \sec\gamma$. A complete revolution of the instrument gives n such equations corresponding to positions at $2\gamma°$ intervals. The resulting displacements Δx cause some variation in the point of contact between the pivot and the probe. However, in view of the small size of Δx_ζ and the smooth changes in the irregularities, this variation may be ignored.

After this, the probe is turned through 180° around the vertical axis. It is now placed, with respect to the vertical line, symmetrically with the original position and will touch the pivot at the point B, forming the same angle γ with the vertical line (see Fig. 24, a).

When the pivot position is such that the point under the probe has the same irregularity $\Delta\rho_\zeta$ the indicator readings may be represented in the form

$$B_{\zeta+2\gamma} = B_0 + \Delta y_{\zeta+2\gamma} - \Delta x_{\zeta+2\gamma} \tan\gamma + \Delta\rho_\zeta \sec\gamma. \tag{9}$$

There will be n such equations corresponding to the number of instrument positions differing by a constant angle 2γ. By subtracting equation (8) from the corresponding equation (9) containing the term $\Delta\rho_\zeta \sec\gamma$, it is possible to eliminate from consideration the interaction between the probe and the pivot, and to construct a system of equations which connect the horizontal with the vertical displacements of the pivot:

$$B_{\zeta+2\gamma} - A_\zeta = (B_0 - A_0) + (\Delta y_{\zeta+2\gamma} - \Delta y_\zeta) + \tan\gamma\,(\Delta x_{\zeta+2\gamma} + \Delta x_\zeta).$$

The derived system consists of n equations and contains $2n + 1$ unknowns: Δx_ζ, Δy_ζ and $(B_0 - A_0)$. In order to obtain an additional system of n equations, the probe with the indicator is placed horizontally, and two similar series of measurements are carried through at the two probe positions C and D (cf. Fig. 24, a).

It is easy to derive equations similar to (8) and (9)

$$C_\zeta = C_0 + \Delta x_\zeta + \Delta y_\zeta \tan\gamma + \Delta\rho_\zeta \sec\gamma,$$
$$D_\zeta + 2\gamma = D_0 + \Delta x_{\zeta+2\gamma} - \Delta y_{\zeta+2\gamma} \tan\gamma + \Delta\rho_\zeta \sec\gamma.$$

Thus, the final system of $2n$ equations for the $2n + 2$ unknowns Δx, Δy and $(B_0 - A_0)$, $(D_0 - C_0)$, is written

$$B_{\zeta+2\gamma} - A_\zeta = (B_0 - A_0) + (\Delta y_{\zeta+2\gamma} - \Delta y_\zeta) + \\ + \tan\gamma\,(\Delta x_{\zeta+2\gamma} + \Delta x_\zeta), \tag{10}$$

and

$$D_{\zeta+2\gamma} - C_\zeta = (D_0 - C_0) + (\Delta x_{\zeta+2\gamma} - \Delta x_\zeta) + \\ + \tan\gamma\,(\Delta y_{\zeta+2\gamma} + \Delta y_\zeta). \tag{11}$$

Imposing the necessary conditions $\sum \Delta x_\zeta = 0$ and $\sum \Delta y_\zeta = 0$, it is possible to determine the unknowns $(B_0 - A_0)$ and $(D_0 - C_0)$ by a simple addition of equations (10) and (11);

$$(B_0 - A_0) = \frac{\sum (B_{\zeta+2\gamma} - A_\zeta)}{n} \text{ and } (D_0 - C_0) = \frac{\sum (D_{\zeta+2\gamma} - C_\zeta)}{n}.$$

Thus, after the elimination of $(B_0 - A_0)$ and $(D_0 - C_0)$ we obtain a new system of equations

$$(\Delta y_{\zeta+2\gamma} - \Delta y_{\zeta}) + \tan\gamma\,(\Delta x_{\zeta+2\gamma} + \Delta x_{\zeta}) = L_{\zeta},$$
$$(\Delta x_{\zeta+2\gamma} - \Delta x_{\zeta}) - \tan\gamma\,(\Delta y_{\zeta+2\gamma} + \Delta y_{\zeta}) = M_{\zeta},$$

where the right hand terms contain the derived quantities and are equal, correspondingly, to

$$L_{\zeta} = (B_{\zeta+2\gamma} - A_{\zeta}) - (B_0 - A_0)$$

and

$$M_{\zeta} = (D_{\zeta+2\gamma} - C_{\zeta}) - (D_0 - C_0).$$

This system is solved by the method of successive approximations. Let us put for the first approximation all $\Delta x_{\zeta}^{(1)} = 0$; then, from the equations $\Delta y_{\zeta+2\gamma} - \Delta y_{\zeta} = L_{\zeta}$ we can find $\Delta y_{\zeta}^{(1)}$ in the first approximation. Correcting for $\Delta y_{\zeta}^{(1)}$ of equation (11), we can find the second approximation $\Delta x_{\zeta}^{(2)}$. Substituting $\Delta x_{\zeta}^{(2)}$ in the equation (10) and solving these equations again, we find $\Delta y_{\zeta}^{(2)}$ for the second approximation. The calculations are continued until the values of the unknown in the succeeding approximations start to coincide. Ordinarily, in view of the small size of tan γ, it is sufficient to carry these calculations only up to the third approximation.

An investigation by this method enables one to estimate the difference between the interaction of the pivot with the probe and with the bearings. For details, see Chapter X.

22. Determination of the Displacement of the Horizontal Axis Resulting from Irregularities of the Pivots

The direct determination of the displacement of the horizontal axis resulting from the irregularities of pivots may be carried out by the method of Airy and by methods of autocollimation.

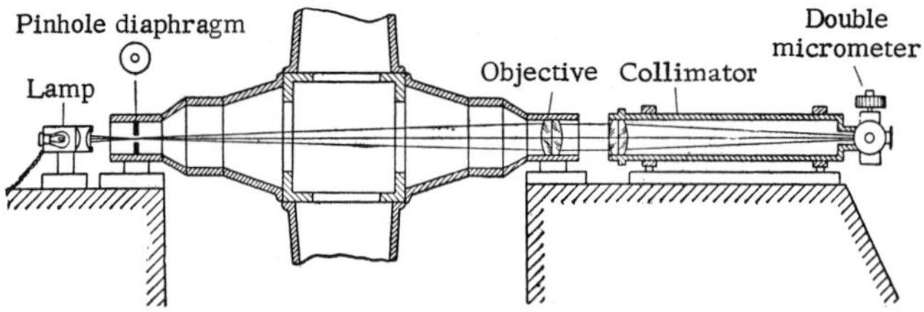

Fig. 25. Assembly for investigating the pivots by the method of Airy.

In the Airy procedure, the hollow axis of the instrument is converted into a special collimator; for this purpose an objective lens is fixed inside the pivot axis in the plane of its working section; the lens has a focal length equal to the distance between the working sections of the pivots. Inside the second pivot is placed a diaphragm with a pinhole, illuminated by a small lamp (Fig. 25). The parallel beam of light coming out of the horizontal axis, converted into a collimator, will change its direction in space during the rotation of the instrument since, together with the horizontal axis, the direction of the line of sight of this collimator is changed. A receiving collimator, focused to infinity and having in the focal plane a double micrometer the screws of which are horizontally and vertically oriented, is placed on a special column opposite the pivot with the lens. The crossed threads of the double micrometer are set on the image of the point source at different elevations of the instrument. The obtained readings x_ζ and y_ζ are related to the unknown quantities Δi_ζ and Δk_ζ through the equations

$$x_\zeta = x_0 + r\sin(m+\zeta) + \Delta i_\zeta,$$
$$y_\zeta = y_0 + r\cos(m+\zeta) + \Delta k_\zeta.$$

These systems of equations are solved analogously to the systems obtained by the Challis method. The method of Airy is superseded by autocollimation methods which are simpler in arrangement and are also more accurate since the latter methods use mirrors.

In the autocollimation method a plain mirror is fastened at one end of the pivot, as perpendicularly as possible to the horizontal axis (Fig. 26). On the opposite end is placed the measuring collimator, focused at infinity and having a Gaussian eyepiece. Then, the rays of light illuminating the set of crosshairs of the double micrometer leave the collimator as a parallel beam, are reflected from the mirror, and form in the focal plane of the collimator the image of the crossed threads. If the horizontal and vertical crosshairs of the micrometer are superimposed on their reflected images, then the line of sight of the collimator, passing through the point of intersection of the two threads, will also be perpendicular to the mirror. The position of this point in the focal plane of the collimator corresponds to the readings x and y . In the case of ideal pivots, when the instrument is rotated the resulting positions of the point describe a circle with a radius which gets proportionally smaller as the angle between the normal to the mirror and the horizontal axis decreases. If the pivots are irregular, they will produce a distorted circle.

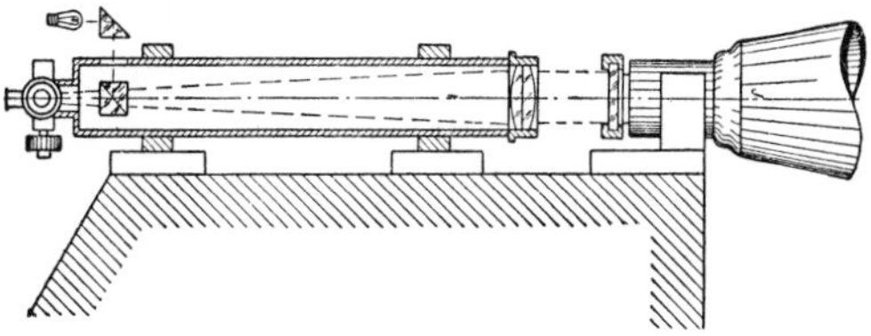

Fig. 26. Autocollimation method of investigation of pivots.

The quantities x_ζ and y_ζ obtainable at different instrument positions are reduced by a method similar to that described in Section 19. The autocollimation method immediately gives changes in the azimuth and the inclination, Δi_ζ and Δk_ζ, due to the irregularities of both pivots.

It is also easy to use photography in the autocollimation method. Instead of the double micrometer, a photographic plate is inserted in a plateholder, and the image of the illuminated diaphragm point source is photographed. The light of the point source is introduced into the collimator by means of a small optical cube (Fig. 26) made of two right-angle prisms cemented together.

Concluding the survey of investigations of the irregularities of pivots, we note that every method has its advantages and disadvantages. The choice of practical methods for investigation is determined by available technical means. It is desirable to investigate the pivots by various methods. The Challis method with photographic recording of the mark and the autocollimation method deserve the greatest attention. The correction of the moments of transit across the meridian by modifications of irregularities of pivots is described in Section 46.

CHAPTER IV

The Instrument Tube

23. The Tube of the Instrument and Its Structure

The tube of meridian instruments consists of two halves: the eyepiece and the objective sections, each resembling a truncated cone with two end flanges. The flanges at the larger ends of the tube halves are securely bolt-fastened to the cube of the instrument, which forms the body of the tube. The objective mounting and a plate for support of the eyepiece micrometer are fastened to the two smaller flanges. The weights of the eyepiece and objective sections of the tube are made equal; hence, if the weights of the eyepiece micrometer and objective are equal, the instrument is balanced with respect to the horizontal axis of rotation. The inside of the tube is painted black and diaphragms are inserted to cut down the scattered light.

The tube is sometimes surrounded by an exterior protecting tube which is also fastened to the cube. The protecting tube is not in contact with the main tube, thereby protecting it against mechanical disturbances as well as against onesided temperature influences which may change the relative positions of the line of sight and the horizontal axis of the instrument. The protecting tube is used for attaching such auxiliary devices as diffraction gratings with their controlling rods, a dewcap with a cover, electric wires, etc. The dewcap, which looks like an extension of the protecting tube for 20–50 cm beyond the objective, is applied in order to protect the lens from fogging at lower temperatures. The dewcap is covered with a lid which protects the objective against dust during the time the instrument is not in use.

The meridian instruments, as a rule, have an achromatic double-lens objective which is corrected visually. In the case where a photoelectric method of recording is used, it must be replaced by a photovisual lens whose chromatic curve largely corresponds to the spectral sensitivity of the photomultiplier.

The best kind of mounting for the objective is a compensating aperture ring within which the lenses are supported by four screws with radial pressure applied at four points along the edges of the lenses. In order to avoid gaps between the lenses and the screws due to the different coefficients of expansion of the metal and the glass, a temperature compensating device is built in which controls the screw length (similar to the one used in pendulum clocks). Along the optical axis the lenses are held against the ring by spring pressure. In the past, lenses were fastened by two springs which pressed the side surfaces of the lenses along two mutually perpendicular radii; these springs fastened the lenses to two surfaces placed inside the internal surface of the aperture ring.

In view of possible shifting of lenses at different elevations of the instrument, this mounting can not be assumed to be completely adequate.

The focusing of meridian instruments is carried out either by inserting gaskets between the flange of the eyepiece micrometer and the flange of the tube, or by a smooth movement of the eyepiece micrometer. During the focusing, one must use large eyepiece magnification. The instrument is considered to be in focus if a sharp star image lies in the plane of the threads of the micrometer.

The most widely accepted method for focusing the instrument is by a smooth movement of the eyepiece micrometer. The eyepiece micrometer is attached to a small tube inserted into a sleeve on the mounting flange. The small tube is displaced with the micrometer by means of a screw which may be securely clamped after each shift. By observing a star image with the eyepiece focused on the crossthreads, and by displacing the entire eyepiece micrometer by the screw adjustment, a position where diffraction rings are clearly visible is obtained. It is convenient to focus the instrument on the surface of the mercury basin to get a sharp reflected image of the crossthreads.

Focusing of the instrument by insertion of gaskets is done less frequently. This method is simpler and more reliable, since no movable eyepiece micrometer mounting is necessary. The gaskets (usually in the shape of disks) may be placed at three or six points between the flanges at the eyepiece end and at the objective end of the tube. In order that they may be used for focusing, it is necessary to determine the thickness of the gaskets. To achieve this, the eyepiece is focused on the star image and either the dioptric reading on the mounting of the eyepiece is noted, or the eyepiece position is marked off by means of an indicator, then the eyepiece is focused on the crossthreads (ordinarily, in view of the fact that the crossthreads are in two different planes, it is sufficient to achieve an average sharpness of the horizontal and vertical crossthreads). From the difference between the two dioptric readings of the eyepiece, or from the amount of linear displacement along the optical axis, it is possible to determine the thickness of the gaskets, which will be required in order to bring the focal image into coincidence with the plane of the crossthreads. A precision of 0.01 mm in gasket thickness is wholly adequate.

In most cases, it is preferable not to change the focal length of the instrument during a series of observations since such a change causes a change in scale. As a rule, changes of the position of the focus are small because their basic causes—change in curvature of the objective and the change of the tube length—compensate each other to a large extent. However, if the temperature variations of the focus make it necessary to change the focusing, then one determines only two constant focusing positions, one for the summer and one for the winter season. This slight lack of focus does not cause a noticeable parallax between the image and the crossthreads.

24. The Illumination of the Field of View and Methods of Decreasing the Brightness of the Stellar Images

Since the crossthreads of the eyepiece are not visible against the night sky, it is necessary to illuminate the field of view. With this illumination, it is possible to observe stars down to the 10th magnitude, depending upon the diameter of the objective and the transparency of the atmosphere. In exceptional cases, the crossthreads are illuminated from the side without illuminating the field of view (whenever it is necessary to observe fainter objects). In this case, the crossthreads become visible against the dark sky in the form of bright lines. The precision of tube alignment is then decreased.

Usually, the field of view is illuminated from the direction of the objective, so that the crossthreads appear dark against the brighter background (spider web threads appear as semi-transparent cylinders and the light passes through their centers; consequently, strictly speaking, they appear as streaks whose edges are darker and their centers brighter). The simplest way to illuminate the field of view is by placing a small lamp at the dewcap in front of the object lens, or inside the tube. Such a method, however, should be avoided since it tends to heat the lens.

The illumination of the field of view through the horizontal axis by a small electric lamp on the pillar of the instrument opposite the pivot opening is a better method. Sometimes, in order to prevent heating the instrument, the lamp is placed at an appreciable distance from it, for instance, on the wall of the pavilion. The use of heat-insulating filters in front of the lamp sharply reduces the possibility of the heat affecting the instrument. In front of the small lamp a light condensing system is sometimes placed. Upon traversing this system the light beam enters the horizontal axis. A diagonal mirror is placed inside the cube of the instrument at an angle of 45° with the optical axis. To avoid obstructing the light path from celestial objects, this mirror has an elliptical opening at its center to admit the light beam from the objective of the instrument. The light of the lamp is reflected in the direction of the eyepiece from the diagonal mirror, thus illuminating the field of view. Sometimes, when there is enough light for illumination, the diagonal mirror may be set so that the light strikes the objective and, upon reflection, illuminates the field of view. In this way, the field of view may be more evenly illuminated.

The intensity of the illuminating beam may be regulated from the eyepiece end of the tube by a rheostat controlling the incandescence of the lamp. Regulation of the brightness of the field of view does not make the stellar image fainter but only changes the contrast between it and the background. In view of the fact that the personal error in observing the transit time of stars depends upon their magnitudes, it is necessary to decrease the apparent brightness of the star images down to the faintest stars of the program.

There are two basic methods for decreasing the brightness of stellar images: by putting a coarse diffraction grating in front of the objective, and by using interchangeable neutral filters in front of the eyepiece. The grating consists of wires stretched in front of the objective at appreciable intervals. After diffraction, the stellar image is weakened, because on either side of it appear the spectral complementary images. The weakening of the central image in comparison with the brightness of the star may be calculated according to the formula

$$\Delta m = 5 \lg\left(\frac{d+l}{l}\right),$$

where d is the thickness of the wires and l is the distance between two neighboring wires. In the case of a cross-wire grating the spectra are spaced along four mutually perpendicular directions, away from the central image. The gratings are put in front of the objective in a rotating frame; ordinarily a set of three gratings is used, reducing the brightness by two, five and eight magnitudes, respectively. One frame opening, with no grating, is used for observations of faint stars. The interchange of gratings is made by a bar which extends along the length of the tube down to the eyepiece.

Fig. 27. Jalousie in front of the objective.

The stepwise decrease in stellar magnitude due to the use of non-variable gratings causes some inconvenience; consequently, a grating with variable interspaces, the so-called jalousie (Fig. 27), is sometimes used. Rotating simultaneously, the parallel strips of the jalousie can, at their extreme positions, either cover the surface of the objective or, when oriented perpendicularly to the surface of the objective, almost completely uncover its surface. The rotation of the strips varies the parameters of the diffraction grating and consequently the brightness of the central stellar image. However, the main cause of the decrease in brightness is the diaphragming of the objective.

As shown by experience, the use of gratings eliminates the so-called magnitude equation. A disadvantage of the gratings and especially of the jalousie is the presence of the complementary spectral images, which interfere with centering of the star image. The rotating frame in front of the objective acts like a sail and causes the instrument to be shaken by a strong wind. A smaller defect is the incomplete use of the objective, which may, generally speaking, lower the quality of the image and change the position of the line of sight. Therefore, it is preferable to use filters at the eyepiece for decreasing the brightness of the star images.

In this method, several (up to seven) smoke-colored filters of various density are placed into a revolving holder in front of the eyepiece. By increasing the brightness of the illumination of the field of view, one decreases the contrast between the intensity of the star image and the background of the field of view. Then the apparent brightness of the image is reduced by introducing the corresponding filter in front of the eyepiece until the image is satisfactory for observation.

The filter chosen for each star is such that the resulting brightness of the star image is comparable with the brightness of faint but still observable stars of the program.

25. Collimation and the Determination of the Collimation-Free Reading

The line of sight of the instrument is in general not perpendicular to the horizontal axis of the instrument, but forms an angle with it of $90° + c$, where c is a quantity called collimation.

In order to determine the collimation it is necessary to know the so-called "collimation-free reading," that is, the reading of the eyepiece micrometer at which the movable vertical thread lies in the plane passing through the second principal point of the objective and is perpendicular to the horizontal axis. The collimation is thus determined as the difference between the collimation-free reading and the reading on the central thread which determines the line of sight of the instrument. In modern instruments, all observations are carried out not on the central thread, which is frequently absent, but with reference to some position in the field of view which corresponds to the mean of the readings recorded as the contacts of the self-registering micrometer.

To determine the collimation-free reading, two horizontal collimators are placed at opposite locations in the pavilion, one to the south and the other to the north of the instrument. The collimators sometimes are placed outside the pavilion. However, a large separation of the collimator from the instrument may blur the crossthread images. This is due to refraction effects caused by the warm air current rising from the ground.

The objectives of the collimators must have diameters at least as large as the diameter of the openings in the central cube of the instrument, and their focal lengths must be approximately equal to the focal length of the lens of the instrument. Each collimator must be provided with a micrometer which may be turned by 90°, or, even better, with a double micrometer. This is necessary for the determination of horizontal flexure (see Section 28).

If the instrument tube is turned to the zenith and both covers in the central cube are opened, by illuminating the crossthreads of one collimator and observing them with the eyepiece of the opposing collimator the two vertical threads can be made to coincide.

After both threads of the collimators have been superimposed, the lines of sight defined by these crossthreads will lie in two parallel planes. These planes define a fixed direction in space, approximately perpendicular to the horizontal axis of the instrument. Since the collimators are focused at infinity, then, provided the telescope tube is directed towards one of them, it is possible to see the image of the crossthreads of the collimator in the field of view of the instrument and set the movable vertical thread of the eyepiece micrometer on the vertical thread of the collimator. Now, turning the instrument 180° about the horizontal axis, one may carry out a similar setting on the vertical thread of the second collimator. The average value of the readings of the eyepiece micrometer, after the settings are made on both southern and northern collimators, will be the reading of the collimation-free position of the movable vertical thread.

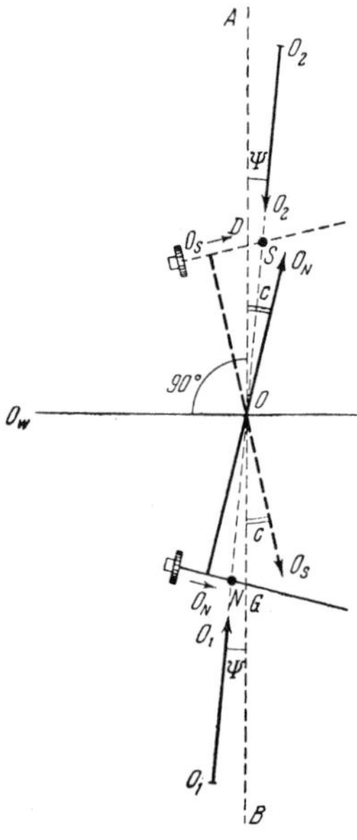

Fig. 28. Determination of collimation.

Suppose that we transpose all the directions parallel to themselves so that they cross at one point, and project the resulting angles on the horizontal plane (Fig. 28). Let O_1O_1 and O_2O_2 be the lines of sight of the southern and northern collimators, subtending an angle Ψ with the line AB which is perpendicular to the horizontal axis O_WO_E. The line AB is the so-called collimation-free line. If the instrument is pointed towards the northern collimator, its line of sight will be O_NO_N and if towards the southern, it will be O_SO_S. The arrows show the directions from the eyepiece towards the objective. The screw of the right ascension micrometer is shown schematically in the directions perpendicular to the lines O_NO_N and O_SO_S. The arrow indicates the direction of increasing readings.

The position of the vertical crossthread of the instrument is shown by the point N when it is set on the vertical crossthread of the northern collimator; let the corresponding reading of the micrometer be m_N. Upon setting the vertical thread on the southern collimator the position of the same thread will be indicated by the point S; the reading of the micrometer will be m_S. Since the angles Ψ and c are in practice small, it is possible to ignore the lack of perpendicularity between the micrometer screw and the direction AB and to assume that $SD = NG$. Therefore the collimation-free reading m_c will be half-way between the readings m_N and m_S, so that $m_c = \frac{1}{2}(m_N + m_S)$. By knowing the middle reading m_0 of the eyepiece micrometer which defines the position of the line of sight of the instrument, it is possible to derive the collimation: $c = m_c - m_0$.

However, since it is impossible to achieve perfect superimposition of two single crossthreads, it is best to provide the collimators with some combination of double crossthreads with different spacing between them. It will then be easy to direct one collimator toward the other and set the narrow double thread inside a wider one. Moreover, it will not be difficult to align the single thread of the eyepiece micrometer with the collimators. Other reticle systems may be used, especially if they are engraved on a glass plate. In order to increase the precision of alignment between the collimators, several trial readings in the alignment process are made, then the movable thread is set on the mean value of these readings. Alignment of the movable thread of the instrument on the collimators is also repeated several times.

The collimation may also be found with only one collimator, by setting the crossthreads of the instrument on the collimator at two different positions of the instrument. However, since resetting the instrument takes some time, there is, as a rule, some danger that during this process of resetting the line of sight of the collimator will change its position in space. Constancy of the principal directions in this case (as well as when two collimators are used) is checked by repeating the measurements with the instrument in the initial position.

To check the collimation and the inclination one may use the mercury basin in the nadir. The instrument is supplied with an autocollimating Gaussian eyepiece and is positioned vertically with the objective above the mercury basis. By superimposing the vertical thread upon its reflected image, we obtain the reading of the eyepiece micrometer, which is nothing else than the collimation-free reading distorted by the inclination of the horizontal axis. If the inclination is determined by a hanging level, then it is possible to eliminate it and to obtain the collimation-free reading. On the other hand, by applying the collimation-free reading obtained with the collimators, it is possible to obtain the inclination. A defect in these testing observations is the fact that the instrument is not in a working position but points downwards.

The frequency of determination of collimation over a given period of observation depends on the rate at which collimation changes with time. There are instruments for which it suffices to measure the collimation once every 10 to 14 days, and sometimes more often—once every 3 to 7 days. With rapidly changing collimation, it is desirable to determine it at the beginning and at the end of each evening.

The variation of collimation with zenith distance is not well known and apparently occurs only in small transit instruments with bent tubes.

26. The Flexure of Meridian Instruments

By flexure we understand deformations in the instrument caused by the force of gravity, that is, deformations displacing various parts of the instrument with respect to each other and changing at various settings of the instrument in altitude. Flexure causes a change in alignment between the line of sight and the chosen zero diameter of the divided circle. Since this change is different at various positions of the instrument, in rotating the instrument around the horizontal axis the line of sight will not turn by an angle equal to that indicated by the divisions of the circle. And conversely, the angle obtained from the readings of the divided circle does not correspond to the angle of rotation of the line of sight. The flexure of meridian instruments, often referred to as *astronomical flexure*, may be divided into tube flexure and circle flexure.

Tube flexure is caused by the deformation of the body of the tube itself, as well as by the displacement and deformation of the objective and eyepiece sections. The change in direction of the line of sight in space is brought about only by a differential displacement of the secondary principal point of the objective with respect to the intersection of the micrometer threads. This difference is sometimes referred to as the differential flexure.

The displacement of the line of sight may be divided into two components, one in the plane of the meridian and the other perpendicular to it. Tube flexure in the plane of the meridian, resulting in a change in the reading of the divided circle with respect to the line of sight, will be discussed in this chapter. The displacement of the line of sight in the direction perpendicular to the meridian plane is called side flexure and will be considered in Section 63.

Circle flexure refers to elastic deformations of the circle, distorting its readings and varying with rotation of the instrument. This flexure is due chiefly to gravitational force, the effect of which is especially strong when the divided circle is not made from a homogeneous material. As a rule, the circle flexure is small and its effect may be reduced somewhat by the use of four reading microscopes.

The correction ΔM_ζ for error in the divided circle reading due to the astronomical flexure may be represented ordinarily by a Fourier series

$$\Delta M_\zeta = a\cos\zeta + b\sin\zeta + a''\cos 2\zeta + \\ + b''\sin 2\zeta + a'''\cos 3\zeta + b'''\sin 3\zeta,$$

where a, b, a'', b'', ... are constant coefficients and the argument ζ is connected with the reading of the circle by the relation $\zeta = M - M_z$, where M is the reading of the circle and M_z is the zenith reading. In this way, the argument ζ changes from 0 to 360° in the direction of increasing circle graduations and is equal to zero at the zenith.

If one neglects the terms above the first-order products, then in the horizontal position of the instrument ΔM_ζ will be equal to $\pm b$, and the coefficient b is therefore called *the horizontal flexure.* The coefficient a is similarly called *the vertical flexure.*

In practice, one considers only those terms with products of the first and second order. This simplification is rather formal, because it is based on the assumption that the size of the terms decreases as the coefficient of ζ increases, owing to the smoothness of the variation of flexure with zenith distance. However, jumps in the variation of flexure are also possible. These sudden changes in flexure may be caused by shifts in the instrument arising instantaneously under the force of gravity, beginning with some instrumental position (for example, a shift in the object lens or the micrometer frame). In that case, the representation of flexure by simple formula is incorrect.

The error due to flexure amounts to $1''$–$2''$ in modern instruments, varying in a complicated way with the argument ζ; it is very difficult to investigate and eliminate completely. Usually the total effect of the tube flexure and the flexure of the divided circle is investigated.

27. Methods of Reducing the Effects of Flexure on Observations

Since the error due to flexure is not determined with certainty, astrometrists are trying to reduce it by organizing their observations in a special way. The following methods will partially eliminate the effect of astronomical flexure.

a) The flexure effect is reduced somewhat by observing each star with two positions of the instrument: the graduated circle first on the east side, then on the west side of the tube.

If the position of the instrument tube at the first observation is I and at the second observation of the same star is II (Fig. 29, a), it is obvious that the corrections for the flexure will be equal for these two positions of the instrument (the arrows in the diagram show the direction of increasing division numbers).

$$\Delta M_\zeta^{\mathrm{I}} = a\cos\zeta + b\sin\zeta + a''\cos 2\zeta + b''\sin 2\zeta + \ldots$$
$$\Delta M_\zeta^{\mathrm{II}} = a\cos(360^\circ - \zeta) + b\sin(360^\circ - \zeta) + a''\cos 2(360^\circ - \zeta) + b''\sin 2(360^\circ - \zeta) + \ldots = a\cos\zeta - b\sin\zeta + a''\cos 2\zeta - b''\sin 2\zeta + \ldots$$

Hence, the zenith distances of stars may be written down as follows:

$$z_{\mathrm{I}} = (M_{\mathrm{I}} + \Delta M_\zeta^{\mathrm{I}}) - M_z, \quad z_{\mathrm{II}} = M_z - (M_{\mathrm{II}} + \Delta M_\zeta^{\mathrm{II}}).$$

The cosine terms drop out if one forms the arithmetical mean of z_{I} and z_{II} or, which is the same, the arithmetical mean of δ_{I} and δ_{II} (they differ from z_{I} and z_{II} only by a constant quantity φ). The mean value of δ will be altered by the following terms, associated with flexure:

$$\Delta M_\zeta = \frac{\Delta M_\zeta^{\mathrm{I}} - \Delta M_\zeta^{\mathrm{II}}}{2} = b\sin\zeta + b''\sin 2\zeta + b'''\sin 3\zeta + b^{\mathrm{IV}}\sin 4\zeta + \ldots.$$

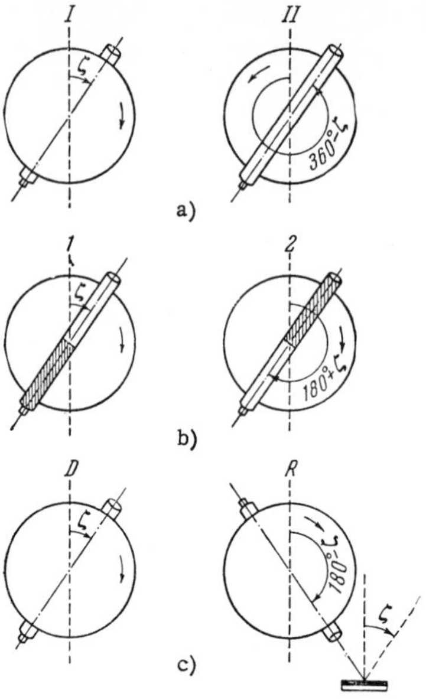

Fig. 29. Reduction of the effects of flexure.

In this way, the average value obtained from the two positions of the instrument does not contain cosine terms; however, the sine terms, including the main term b sin ζ, are not eliminated.

b) The effect of flexure may also be decreased by repeating the observations after interchanging positions of the objective portion and the eyepiece portion. To be able to do this it will be necessary that the eyepiece portion (with micrometer) and the objective portion (with its lenses) have equal weights and that the end flanges of the objective and eyepiece portions of the tube are similar. Thus, a fairly easy interchange of these portions will be possible, without disturbing the balance of the instrument.

Suppose that for the first relative position of the objective and eyepiece portions the correction due to flexure is

$$\Delta M'_{\zeta} = a \cos\zeta + b \sin\zeta + a'' \cos 2\zeta + b'' \sin 2\zeta + \ldots$$

Upon interchanging the objective and eyepiece sections the correction due to flexure will be

$$\Delta M''_{\zeta} = -a \cos\zeta - b \sin\zeta + a'' \cos 2\zeta + b'' \sin 2\zeta + \ldots,$$

provided other characteristics of the instrument remain unchanged. The quantity ζ changes by 180° (Fig. 29, b) when the instrument is set on the same star in the second position. The corresponding zenith distances will be

$$z_1 = (M_1 + \Delta M'_{\zeta}) - M_z, \quad z_2 = (M_2 + \Delta M''_{\zeta}) - M_z - 180^\circ.$$

Thus, when the average of the zenith distances is taken, the following terms associated with flexure are retained:

$$\frac{\Delta M'_\zeta + \Delta M''_\zeta}{2} =$$
$$= a'' \cos 2\zeta + b'' \sin 2\zeta + a^{IV} \cos 4\zeta + b^{IV} \sin 4\zeta + \ldots.$$

They contain sine and cosine terms of even multiples of angles. Forming the half-differences of the readings, we obtain

$$\frac{1}{2}[(M_2 - 180^\circ) - M_1] = \frac{1}{2}(\Delta M'_\zeta - \Delta M''_\zeta) =$$
$$= a \cos \zeta + b \sin \zeta + a''' \cos 3\zeta + b''' \sin 3\zeta + \ldots,$$

and we determine the coefficients a, b, a''', b''', ...

The effect of flexure on accuracy of observation can be reduced significantly by interchanging the eyepiece and objective portions and performing observations at two positions of the instrument. In this way, we retain only sine terms with even multiples of angles, that is, the terms with the coefficients b'', b^{IV}....

However, it should be noted that the interchange of the eyepiece and the objective will eliminate the odd terms in the expression for the flexure under the condition that only the two halves of the tube and the cube are deformed and not the objective and the eyepiece parts. Otherwise some unaccounted-for flexure of the tube will remain and will depend upon the deformation and displacement in the objective and the eyepiece micrometer as well as upon how tightly the screws which hold them are fastened to the tube.

The described methods of observation a) and b) are frequently used in practical astronomy for constructing the absolute catalogs of stellar positions.

c) The method of observation utilizing the stellar images reflected from a mercury surface and thus eliminating the necessity of determining the zenith point was suggested by Bessel. It was used at several observatories, particularly for long series of observations carried out at the Greenwich and Washington observatories.

In this method, the direct observations of stars are combined with the observations of the reflected images obtained with a mercury surface placed in the meridian of the instrument. A basin of mercury is placed on a stand so that by pointing the objective section of the tube downwards, at an angle of 180°-z with the plumb line, it is possible to see the star having a zenith distance z (Fig. 29, e). To observe stars at different declinations, the basin with mercury must be shifted in height or moved along the meridian.

This method of observation reduces the effect of flexure. Actually, the directly observed zenith distances z_D and the values z_R obtained from observations of reflected star images may be represented by the following formulas:

$$z_D = (M_D + \Delta M^D_\zeta) - M_z, \quad z_R = 180^\circ - [(M_R + \Delta M^R_\zeta) - M_z].$$

The arithmetical mean of these quantities will contain neither the odd terms containing sines, nor the even terms containing the cosines:

$$\Delta M_\zeta^D = a\cos\zeta + b\sin\zeta + a''\cos 2\zeta + b''\sin 2\zeta + \ldots,$$
$$\Delta M_\zeta^R = -a\cos\zeta + b\sin\zeta + a''\cos 2\zeta - b''\sin 2\zeta + \ldots,$$

since the argument ζ is equal to 180°-ζ in the second case.

However, when this method of observation was used it was found that the systematic differences $z_D - z_R$ do not agree with the flexure determined by other means. This, obviously, is explained by a different behavior of the instrument when the tube points downwards. This last problem and the small range of inclinations available for such observations, as well as the considerable inconveniences connected with these observations, prevented a wide use of this method.

28. Laboratory Methods for Determining the Flexure

Since the methods of observation presented in the previous section only partially eliminate the flexure errors from the observed zenith distances, it is necessary to determine the flexure by some special methods. The most frequently employed methods will be presented now, chosen out of the many different laboratory methods. These, as far as possible, will represent a variety of procedures.

a) The most widespread method is the method for determining the horizontal flexure with the collimators, one northern and one southern, as suggested by Bessel. In this method, each collimator must have a micrometer with a movable horizontal thread.

The collimators are set so that their lines of sight are parallel to some plane as near as possible to the horizontal plane. This is achieved by making the horizontal threads of the collimator micrometers overlap as is done in determining the collimation. The horizontal crossthreads of the instrument are then set on those of the collimators and the eyepiece micrometer and divided circle readings are taken. If flexure exists, the readings will be distorted and will differ by 180° only after correcting for flexure:

$$M_S + \Delta M_\zeta^S = (M_N + \Delta M_\zeta^N) \pm 180°,$$

where M_S, M_N, ΔM_ζ^S and ΔM_ζ^N are the circle readings obtained with alignment on the southern and northern collimators and their respective corrections for flexure.

Let us assume that the argument ζ is equal to 90° when the instrument is set on the southern collimator. Then, when the instrument is directed towards the northern collimator, ζ will be 270° and

$$\begin{aligned}(M_S - M_N) \mp 180° &= \Delta M_\zeta^N - \Delta M_\zeta^S = \\ &= (a\cos 90° + b\sin 90° + a''\cos 180° + b''\sin 180° + \ldots) - \\ &- (a\cos 270° + b\sin 270° + a''\cos 540° + b''\sin 540° + \ldots) = \\ &= 2b - 2b''' + 2b^{V} + \ldots\end{aligned}$$

If we assume that the coefficients of terms which contain the sines of angles with even multiples larger than two are small, then

$$\tfrac{1}{2}(M_S - M_N) \pm 90° = b.$$

By investigating the variations of the horizontal flexure at different positions of the limb on the axis of the instrument, it is possible to determine the flexure of the divided circle.

The vertical flexure a is also determined in a similar manner; for this purpose, the so-called vertical collimator is used as well as the mercury surface at the nadir. The vertical collimator is placed horizontally on one of the stands of the apparatus. A plane mirror is placed in front of the objective of the collimator at an angle of 45° to its axis, permitting alignment of the instrument on the collimator's crossthreads. The vertical collimator and the instrument are supplied with autocollimating eyepieces. To begin with, the vertical collimator must be placed along the required direction. For this purpose, with the instrument at the horizontal position and the aperture of the cube uncovered, the real crossthreads of the collimator are superimposed on their images reflected from the mercury surface. This fixes in space a plane parallel to the plumb line and close to the plane of the prime vertical. Then two readings of the eyepiece micrometer and of the divided circle are taken: one, by setting the horizontal movable threads of the instrument on the threads of the vertical collimator when the tube is pointing toward the zenith; the other, by superimposing the threads of the instrument with their images reflected from the mercury surface when the tube is pointing at the nadir.

Similarly, for the determination of horizontal flexure we obtain:

$$(M_c - M_s) \pm 180° = 2a + 2a''' + 2a^V + \ldots,$$

where M_c and M_s are the instrument readings when the instrument points toward the vertical collimator and the mercury surface, respectively; if one ignores the higher order terms in the series, we have:

$$\frac{1}{2}(M_c - M_s) \pm 90° = a.$$

b) In the method of Elistratov the use of collimators permits determination of flexure for any zenith distance and, with slight procedural complication, permits separate determination of the tube and circle flexures.

Collimators with Gaussian eyepieces are placed at different zenith distances above the instrument in the plane of the meridian and along a circular arch (Fig. 30). A mirror is attached to the side of the central cube of the instrument. The perpendicular to the mirror makes an angle of 90° with the line of sight, and is close to the plane described by the line of sight.

Let us assume that the instrument is placed at such a position that the mirror is nearly perpendicular to the optical axis of the collimator, which is placed at some zenith distance z_i. Let us set the horizontal thread of the collimator on its reflected image and note the reading of the circle M_i'. After this, we rotate the instrument through 90° and point it at the collimator; let us now set the working thread of the instrument on the horizontal thread of the collimator and note again the circle reading M_i. The setting of threads in either position is achieved

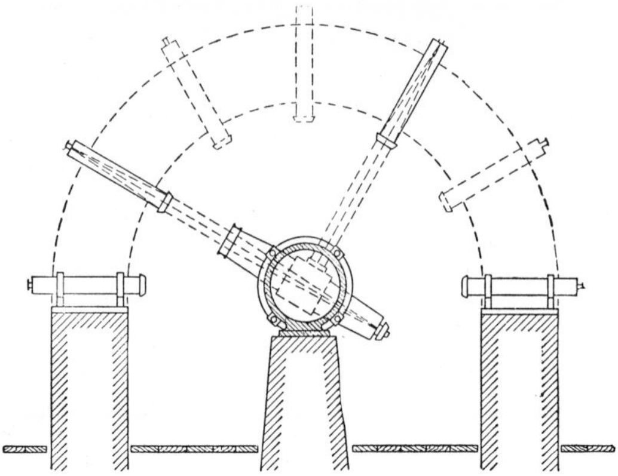

Fig. 30. Determination of flexure according to Elistratov.

by a micrometric motion of the instrument tube so that the readings of the circle are made on the same graduations of the limb. Such measurements may be carried out at different zenith distances z_i.

Let x_i be the correction for flexure of the tube and y_i the correction for flexure of the circle and for its torsion with respect to the central cube. When these corrections are applied to the difference of circle readings made in two positions of the instrument it will differ from 90° by a small angle ι due to the inaccurate mounting of the mirror on the cube:

$$M_i' + y_i' - M_i - y_i - x_i = 90° - \iota$$

or

$$M_i' - M_i = 90° - \iota + x_i + y_i - y_i'. \tag{12}$$

We should note that the use of four circle microscopes placed at 90° intervals will eliminate the graduation errors of the circle from this difference.

The flexure of the circle may be determined if a second mirror is fastened to the opposite side of the cube at a certain angle ω with respect to the first mirror. Let M_i'' be the circle reading when the second mirror is facing the collimator. Then, the two observations of the images of the threads reflected by the two mirrors give the relation $M_i'' + y_i'' - M_i' - y_i' = \omega$, where ω, the angle between the mirrors is assumed to be constant. If one resets the collimator to the zenith distance $z_{i+1} = z_i + \omega$, then one can see that $y_{i+1}' = y_i''$.

Choosing ω so that $n\omega = 360° \cdot k$ where k and n are integers, we obtain a system of equations for determinating y_i' which is similar to the system for determining periodic errors by the method of Rydberg:

$$y_{i+1}' - y_i' = M_i' - M_i'' + \omega.$$

However, these equations will be closed only if the collimators are placed along the whole length of the circle; this is, however, technically impossible for negative altitudes. Consequently, one should use two collimators pointing at each other in the horizontal plane, as in the Bessel method. Then all the equations remain unchanged except the last one, which takes the form

$$y_1' - y_n' = M_n' - M_1' + 180^\circ.$$

After determining the quantity y', it may be applied in equations (12) according to the corresponding arguments M and M' to obtain the true tube flexure. By cutting each mirror in two halves and attaching these two to the central cube separately one may check the required constancy of the mirror positions. Then two reflected images of the crossthreads will appear in the field of view of the collimator. The degree of steadiness of the distance between the two images will serve as a criterion of the absence of deformations of the central cube and of the mirrors.

The disadvantage of this method is the necessary complicated structure of the arched frame for attaching the collimators and the inconvenience of using them.

c) The method of Bonsdorf for determining the tube flexure was proposed and tested at the Pulkovo Observatory during the investigation of the vertical circle of Ertel. Two rods AE are placed along the instrument tube and fastened as tightly as possible to the free sides of the central cube. They support a mirror C on a special mounting in front of the objective (Fig. 31). The displacement of the autocollimating image of the horizontal threads during the change in height of the instrument position is caused by the tube flexure and may be directly measured by superimposing the crossthreads on their reflected images. The mirror must be fastened in such a way that the direction of the perpendicular to the mirror does not change as a consequence of the flexure of the instrument. When this is done, the angle of rotation of the perpendicular to the mirror (when the instrument is rotated) will correspond to the angle described by the pivot around the horizontal axis. For this purpose the transverse crossbeam D is connected to the rods with the hinges E. Thus, if the rods are bent, the mirror will be displaced only in the direction parallel to itself. This was confirmed experimentally in the following way: the autocollimating image was not deflected when a 5 kg weight was attached to the ends of the rods with the instrument in the horizontal position (this is a relatively large weight since the total weight of the instrument is about 9 kg). This method is applied with the necessary assumptions that any distortion of the central cube is independent of the zenith distance and that tube flexure is not altered by attachment of equipment to the cube.

d) The Lévy method involves the introduction of different auxiliary optical systems into the central cube in order to determine the flexure of the tube. The use of these optical systems always entails the doubtful assumption that there is no flexure within the auxiliary system itself. This is the reason why they are not used very often. However, the method proposed by Lévy seems very interesting and a short description of it is given below.

Within the cube of the instrument there is placed a convex-concave lens whose focal length is one half the focal length of the tube. Thus,

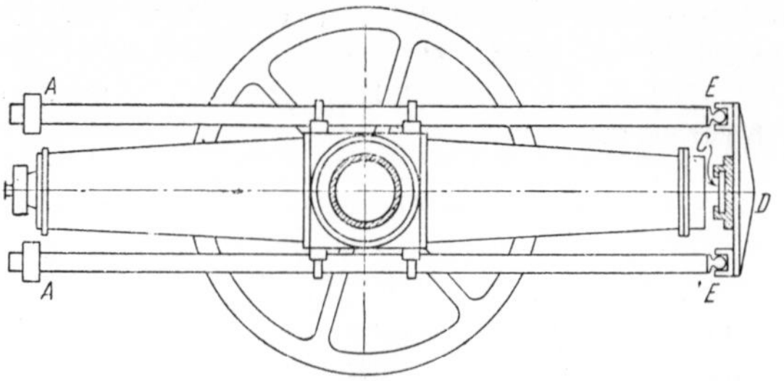

Fig. 31. Arrangement of Bonsdorf for investigating telescope flexure.

a mark on one of the lenses of the objective will be seen in the field of view (Fig. 32). The concave surface of the inserted lens faces the eyepiece and is silvered with the exception of two portions A and B, one at the center and the other at the edge of the lens. The curvature of the concave surface is such that it forms a reflected autocollimating image of the eyepiece crossthreads in the field of view of the instrument.

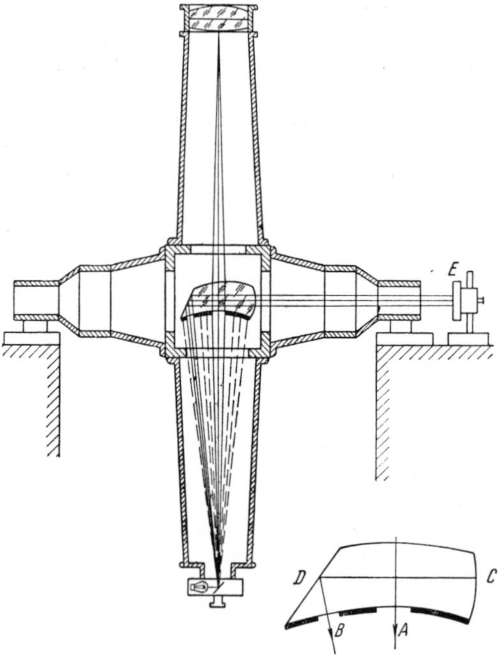

Fig. 32. The use of the Lévy lens.

The displacement of the image of the mark on the objective lens is observed through the central portion of the concave mirror at various elevations. The measurement of this displacement in declination by the eyepiece micrometer will give us the total effect of the deformation of both halves of the tube, including the objective lens and the eyepiece micrometer. By measuring the shift of the crossthread images reflected by the concave mirror, we obtain double the displacement due to the deformation of the eyepiece half of the tube only.

The difference between these two displacements (that of the image of the mark on the objective and that of the autocollimation image of the eyepiece micrometer thread) is equal to the difference between the deformation of the objective and the eyepiece halves of the instrument tube, i.e., the flexure of the tube.

This lens may also be used in investigating the effect of the pivot irregularities on the position of the horizontal axis. For this purpose, one side surface of the lens is made convex C, and the other is plane and at some angle to the horizontal axis D. On the pillar of the instrument, in front of the pivot, is placed a mirror E which reverses the light beam from the autocollimating eyepiece and sends it through the unsilvered side B of the lens. Due to the irregularities of the pivots, the position of the lens with respect to the mirror will vary. This will produce a displacement of the images of the autocollimating crossthreads, and its measurement will determine the irregularities of the pivots.

This concludes the description of methods of investigating the flexure of meridian instruments; however, one should keep in mind that the problem of determining the astronomical flexure can not be considered solved. The laboratory methods for determining the flexure are rather numerous, and to a large extent not tested in practice. Therefore, it is difficult at the present time to recommend any of them.

CHAPTER V

The Graduated Circle

29. Graduation of the Circles

The division marks of graduated circles are engraved on a metal band pressed along the edge of the circle into a groove whose cross section has the form of a swallowtail. This band with its division marks is called the limb of the graduated circle. Sometimes the word limb is used to designate the whole graduated circle.

The limbs of graduated circles are made of various materials, the most frequently used being silver, German silver (a copper-zinc-nickel alloy) and a platinum-iridium alloy. The limb material must be resistant to corrosion and neither too soft (lest the division marks be obliterated by repeated cleaning) nor too hard (lest the cutting tool become blunted before engraving of the division marks is finished). In geodesy, glass graduated circles are often used. However, among astrometric instruments, only one meridian circle is equipped with glass circles, and that one is at Greenwich Observatory in Herstmonceux. The difference in the temperature coefficients of expansion between the glass and the material of the horizontal axis (brass, steel) makes the fitting of the glass circle on the horizontal axis complicated.

Circles of astrometric instruments have divisions for every two or five minutes of arc. The modern instruments more frequently have divisions for every five minutes. The division marks are set along the radii of the circle, and, if prolonged, would intersect at a point referred to as the center of division.

The divisions may be engraved in two ways: by copying a master circle or through automatic graduation. During the copying, the blank circle is placed on the same axis with the master circle and fastened to it. By rotating both circles simultaneously, one sets the division mark of the master circle between the double threads of the microscope which controls the engraving tool. Then the engraving tool is moved and cuts a line on the blank circle. When the tool is set with the microscope, the errors of the master circle should be taken into account. Therefore, the master circle must be carefully investigated in advance. After the division mark has been engraved on the blank circle, the latter, together with the master circle, is rotated until the next division mark to be copied enters the field of view of the microscope. The circle is sometimes copied in sections by applying a curvilinear scale between the previously engraved division marks.

This method of copying was widely used in the last century. The limbs were produced in this manner by the well-known firm of Repsold. It is to be expected that those errors in the master circle which are not completely taken into account during the duplicating are transferred to the copy. This is easily seen by looking at Fig. 33, which represents by smooth curves the errors of the diameters of several circles. Nowadays, this method of graduation has only historical interest and is completely superseded by the automatic method of graduation.

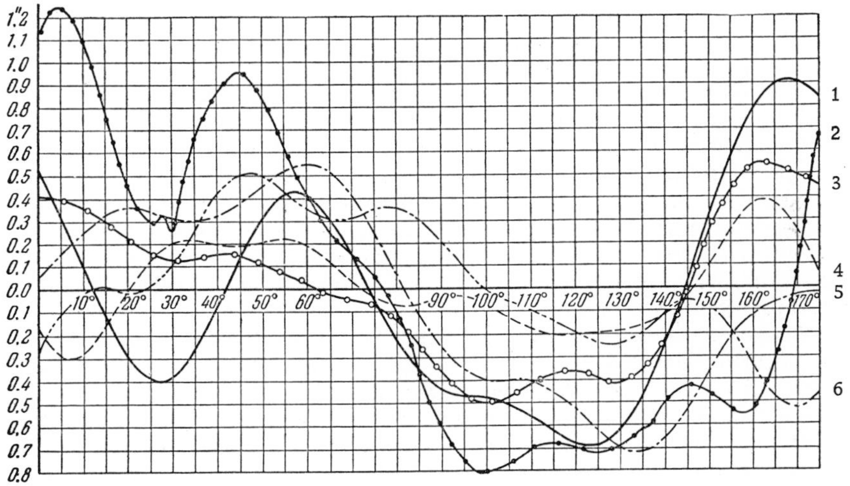

Fig. 33. Corrections of the diameters of Repsold circles of different observatories. 1—Tashkent; 2—Kharkov; 3—Pulkovo; 4—Kazan; 5—Moscow; 6—Odessa.

The automatic graduating machine consists of a large precision-made gear set horizontally. A worm screw engaged with the gear can be turned through definite, predetermined angles. After the gear together with the axis of the blank circle is rotated, the mechanism of the engraving tool is activated and a division mark is made. The cycle is then repeated.

The division errors in the automatic graduation method may arise from the inaccuracy of the large gear and of the worm screw; from inadequate stability of the driving gear, the rotating gear, and the movable engraving tool; and, finally, from the instability of the machine as a whole. It is very important that after each cycle the moving parts of the machine resume their previous relative positions, otherwise short-periodic division errors may appear.

The graduating machine must be put on a separate foundation, isolated from the building, so that it is not subject to jarring. Since the process of graduation of the limb sometimes lasts many hours, the temperature of the premises in which the graduating machine is located must be constant or the machine must be placed in a thermostatically controlled booth.

After the graduation is finished, the limb is cleaned with pumice in order to remove all the rough edges of the division marks. Then the limb is polished and the engravings are filled with paint or carbon black. The division marks are numbered in no specified way since this is necessary only for identifying individual marks. Ordinarily, during observations, whole degrees are not recorded, and in processing the data any convenient numbering is assumed.

30. Reading the Graduated Circle with Microscopes

The graduated circles of astrometric instruments are read off with the help of reading microscopes equipped with micrometers for visual reading, and with photocameras for photographic recording. A photoelectric method of reading the graduated circle has been devised recently.

The principal parts of the visual reading microscope are the objective, the micrometer, and the eyepiece. The small microscope tube holding these parts together is fastened to a special stand on which it is possible to adjust the position of the optical axis of the microscope. The problem of attaching and adjusting the reading microscope will be considered later; for the present we shall assume that its optical axis is perpendicular to the plane of the graduated circle and that the distance of this axis from the center of the circle is the same as that of the midpoint of a division mark.

The objective of the reading microscope (ordinarily consisting of two lenses) forms a magnified image in the plane of the micrometer threads. Objectives with long focal length are generally more convenient since the longer microscopes have greater relative stability. A displacement of one of the ends of the microscope in this case causes smaller angular tilting of its optical axis. The length of the reading microscopes of different instruments ranges from 150 to 600 mm. The objective magnification of the microscope does not ordinarily exceed five. The micrometer threads and the images of the circle divisions are usually viewed through an eyepiece with magnification of 6 to 10 in order to increase the accuracy of the settings.

The filar micrometer is placed inside a small box fastened to the eyepiece end of the reading microscope. The micrometer screw, mounted inside the box, moves the small frame with crossthreads. The reading microscopes are placed in such a way that the distribution of the division marks of the circle with respect to the threads will be the same in the field of view of each microscope and their graduated head readings will be closely similar.

The spider web threads of the micrometer are stretched in the form of a close pair with the distance between them from two to three times larger than the image of a division mark. The setting on a division mark is done by a positive turn of the micrometer screw until the interspaces between the threads and the division mark become equal on either side of it. The setting of a double thread on a division mark is many times more accurate than the setting of a single thread.

The objective magnification of the reading microscope usually conforms to the pitch of the micrometer screw so that one complete revolution of the screw will correspond to a displacement of one minute of arc against the marks on the limb. The value of one division on the screw head (divided into 60 parts) will then correspond to one minute of arc on the limb. Therefore, the setting of a double thread may be made with an accuracy down to 0.1 of a second. The setting precision depends on the quality of the division marks and on the microscope magnification. The root mean square error of a setting is about $\pm 0''.25$.

The micrometer reading is made up from the denomination of the lower division mark of the two (the one which is closer to the crossthreads), from the number of minutes read off the notched scale of the micrometer, and, finally, from the number of seconds and tenths of seconds obtained from reading the graduated head of the micrometer screw. Sometimes, in order to increase the accuracy and decrease the possibility of personal errors and errors of the division of the limb, settings are made both on the higher and lower numbered division mark and the average of the readings of the micrometer head is taken as the final one.

Everything presented above is valid as long as the scale of the reading microscope corresponds exactly to the calibration of the screw head, that is, k rotations of the micrometer screw correspond to the interval between two neighboring division marks on the limb. If this is not the case, then it is necessary to calculate the scale error, sometimes referred to as the run. The run is conventionally defined as the difference between the two micrometer head readings corresponding to settings on the higher and the lower numbered division marks: $r = b - a$. When settings are made on both neighboring division marks, the run correction is equal to

$$\Delta M_r = \left(\frac{a+b}{2}\right) \cdot \frac{r}{60 \cdot k} - \frac{r}{2}. \tag{13}$$

Sometimes two double threads are stretched in the field of view with a distance between them of 0.5 to 1.5 screw revolutions. One of them is set on the higher numbered and the other on the lower numbered division mark. In that case, in the formula (13), $\frac{a+b \pm 30''}{2}$ is substituted for $\frac{a+b}{2}$.

In order to minimize the effect of accidental errors of the division marks on the determination of the run, it is derived for each reading microscope by special measurements at various sections of the circle. Systematic variation of the run along the circle would indicate that the graduated circle is not perpendicular to the axis of rotation of the instrument.

To correct the observations for the run, the magnification of the reading microscope is adjusted by changing the distance between the object lens and the plane of the crossthreads. After the object lens is shifted, the whole reading microscope naturally has to be focused again on the division marks of the circle. In practice, the reading microscopes are adjusted by a series of tests.

Since crossthreads engraved on a glass plate are not as yet widely used, we shall describe briefly the technique of mounting the spider web threads.

The spider web threads are obtained from a spider's cocoon, found in the spring, and put into boiling water in order to kill the larvae. The cocoon threads have a true cylindrical shape. The thickness of the threads is constant along their length and varies on the average between 4 and 10 microns. Extra long spider threads are stretched by attaching small lead weights of one to two grams to their ends, depending upon their strength and thickness. With a knife a notch is cut in each weight, the thread is inserted into the notch and is fastened there by pressing the weight with a pair of pliers. It is most convenient to cut the thread from the cocoon after the second weight is attached. It is advisable to prepare extra threads since they may break during the procedures that follow. When stretching out the threads, it is convenient to work against a black background and to illuminate the threads with a sidelight from a table lamp.

As a rule, there must be notches on the micrometer frame for placing the threads. In case there are no notches, one must cut them with a dividing machine. Before the threads are fastened, the notches must be cleaned from old paint, glue, or grease. This is done with alcohol, benzine, or ether, and by using small wooden sticks and cotton in order not to damage the notches.

The spider web thread is now immersed for several minutes in boiling water so that the thread becomes thinner; this is done by holding it by the weight with tweezers. The threads are fastened afterwards to the notches of the frame, which is taken from the micrometer and placed on a high stand. If the threads do not fit directly into the notches, then the thread ends, lying on the frame, are in turn lifted from their places by a needle, and by observing them through a strong magnifying glass, are placed into every notch separately. With the threads correctly placed, the weights must hang freely.

After the threads are placed, they are immediately glued on. One may use liquid shellac or some appropriate glue; the latter, if needed, may be diluted with alcohol. The glue is applied in thin layers with a needle. It is useful to dry the frame with the threads in a humid atmosphere by placing it under a glass cover together with hot water. In twenty-four hours the excess ends of the threads with the weights attached to them are cut off with a razor blade, and the frame is replaced in the micrometer.

31. The Mounting of Reading Microscopes

The reading of graduated circles of astrometric instruments is done with four microscopes. To eliminate the eccentricity error, two microscopes are sufficient. However, the use of four microscopes is usually preferred in order to help eliminate the lack of perpendicularity of the plane of the limb to the horizontal axis, to reduce the effect of circle flexure as well as of systematic and accidental graduation errors, and to reduce the accidental reading errors.

The microscopes are mounted on a special frame or, what is even better, on a circular drum so that the diameters joining them form 45° angles with the horizontal. If the readings are made with two microscopes, they are spaced on the top of the supporting pillars of the instrument. The microscope frames of the meridian circles are suspended on the inner surface of the supporting pillars, ordinarily on the same flange that supports the bearings. However, this method is inferior to the method in which the microscopes are mounted on drums which are placed, together with the bearings, on the top surfaces of the supporting pillars. This assures a greater stability of the microscopes and even of the instrument. Drum mounting of the microscopes also has the advantage that the angles between them can be changed. This is necessary for investigating the divisions of the limb and is not usually possible if the microscopes are mounted on a frame.

The difference between the circle readings must correspond to the angle of rotation of the instrument. For this to be correct, the drum must be centered and the reading microscopes must be placed properly.

The centering of the frame or of the drum with the reading microscopes is done so that the zero points of any two opposing microscopes are at the ends of one diameter; moreover, the two diameters formed by the two pairs of opposing microscopes should be mutually perpendicular and their point of intersection should coincide with the axis of rotation of the graduated circle.

The drums with the microscopes must be adjustable so that the azimuth of the axis of the drum and its inclination in the plane of the prime vertical may be changed. It is not necessary to rotate the drum around its axis since this adjustment may be made by moving the microscopes along the rim of the drum. Moreover, it must be possible to move the drum along the meridian and in the vertical direction. All these adjustments are accomplished with the corresponding screws, which operate by shifting the flat base which is an integral part of the drum.

The correct position of the drum and the reading microscopes may be located by testing, that is, by making successive readings at different circle positions after each adjustment.

The method of Bessel is used to verify the correct position of the drum and the microscopes. If the measurements are carried out by this method, it is possible to estimate the size of the errors of the pivots and of the circle graduations.

When the microscopes are mounted on the drum, one must be sure that each microscope may be moved on the drum radially and tangentially. Moreover, it must be possible to change the inclination of the microscope in the plane perpendicular to the direction of the marks, and to focus the microscopes on the marks.

To achieve this, the microscopes are placed in two ring holders with slits which may be tightened by screws. This makes the focusing of the microscopes very simple. The ring holders are mounted on small flanges which may be fastened to the drum rim or to the frame. By moving the flanges, one may move the microscopes tangentially. The remaining adjustments are achieved by regulating the position of the ring holders with respect to the flanges.

Lamps for illumination of the limb sections under the microscopes are attached to the drum edge facing toward the graduated circle. The limb should not be illuminated from the side, since shadows would affect the observed position of the division marks. For the same reason, the stability of illumination is also important. The lamp is placed along the same radius as the microscope, with the light beam striking the limb along the division mark at a certain angle to the microscope axis. Sometimes, with the lamp placed on one side of the microscope, the light beam is directed along the microscope axis by means of a plaster or paper reflector placed at 45° to the axis and having a central aperture. In the case of visual readings, it is convenient to use green filters.

32. Photographic and Photoelectric Methods of Reading the Circle

The photographic recording of circle readings is now used to some extent. Cameras for photographing a section of the limb are ordinarily placed on each microscope. A divided optical system (optical cube) also makes it possible to take visual readings. The images of the four limb sections are sometimes brought together on one frame. In that case, the camera is placed at the center of the drum. Figure 34 shows the common camera for the four microscopes as well as the cameras on the supplementary microscopes. These supplementary microscopes are used in investigating division errors of the graduated circle.

Fig. 34. The drum with photomicroscopes.

For photographic recording of the circle readings, the microscope must have a fixed index in its focal plane parallel to the images of the division marks and photographed simultaneously with them. Ordinarily, a stretched spider web thread or a line on a glass plate is used as the index. After pressing the release button, the shutter of the camera is automatically opened and the illuminating lamp lighted for the exposure time. After the exposure, the shutter closes and the film is moved on. The microscope scale must be such that, on any single photograph, at least two division marks of the limb are visible. The view of the film with images of the limb and the index is shown in Fig. 35. Sometimes the images of supplementary threads are obtained on both sides of the index image; these serve as additional reference lines. To obtain the circle reading, it is necessary to measure the distance between the index and the nearest mark and to express it as a fraction of the interval between two neighboring division marks. This may be done by also measuring the distance between the images of the two neighboring division marks of the limb. If the scale of the photographs remains unchanged, it is sufficient that these supplementary measurements be carried out only on a few of the frames at the beginning and at the end of the observing period. Measurement of the division mark images has a mean square error of ±2–3 microns and of the index image ±1–2 microns. If the distance between the images of two five-minute division marks on the film is 10 mm, then the general precision of a photographic recording is approximately $\pm 0''.10$–$0''.20$; that is, it will be roughly equal to the precision of visual circle reading.

Fig. 35. Photograph of a portion of the limb.

Photographic recording of circle readings cuts the time necessary for observing one star; it records the circle reading almost simultaneously with setting the threads of the eyepiece micrometer on the star image. However, measurement of the photographic film is a time-consuming process (efforts are being made to render the process automatic).

A completely automatic recording of the circle readings is possible if one uses photoelectric reading microscopes. Recently, several models of photoelectric microscopes have been developed and used successfully in measuring machines for measuring spectrograms and also to record the readings of graduated circles and scales. We present a brief description of one of the structural arrangements for a photoelectric reading microscope.

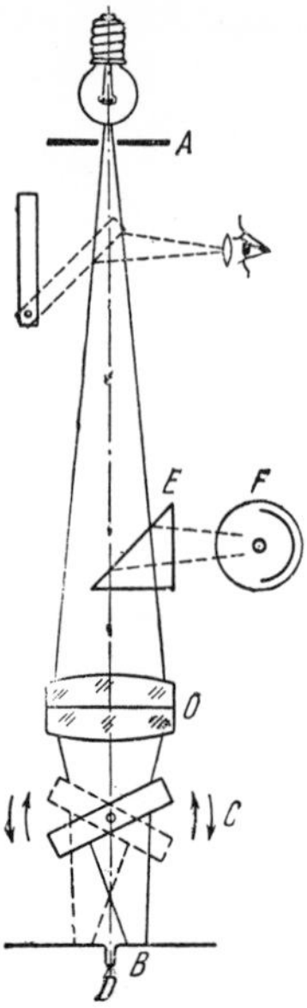

Fig. 36. Schematic of the photoelectric microscope.

The objective O of the microscope (Fig. 36) forms an image of the slit A (illuminated by a small lamp) on the surface of the limb; this image is a small bright strip B parallel to the division marks. A plane-parallel glass plate C is vibrating with a frequency of 50 cycles, so that the bright strip makes 50 complete vibrations per second with an amplitude of about 5 microns, crossing the measured division mark D in 1/50 of a second. The light flux reflected from the surface of the limb passes through prism E and strikes the photoelement surface F. The electric current generated in the circuit of the photoelement will have minima at the moments when the bright strip and the division mark coincide. The electronic circuit transforms this current into short electric pulses which correspond to the minima of the current. The pulses will have a phase difference of 180° only when the division mark is at the midpoint between the two extreme positions of the bright strip. The microscope can be moved with a micrometer screw rotated by a motor. The phase difference of the pulses controls the motor until the difference is equal to 180°. The motor is then stopped and the reading of the graduated head of the micrometer screw is photographed. The setting accuracy of such a microscope is about ±0.25–50 microns which, for a circle of 100 cm diameter, corresponds to $\pm 0''.10 - 0''.20$.

33. Misalignment of the Graduated Circle Center with the Instrument's Axis of Rotation

In attaching the graduated circle to the horizontal axis the center of the circle should coincide with the axis of rotation of the instrument.

However, this is very difficult to achieve. If it is not done, the angle obtained from the readings of one microscope will not be equal to the actual angle of rotation of the instrument around the axis. The difference between the two angles is called the eccentricity error.

Let us designate the axis of rotation by O, and the graduation center by O' (Fig. 37). The distance OO', expressed in microns, is called the *linear eccentricity*; when expressed as a fraction of the radius of the graduated circle and stated in seconds of arc it is called the *eccentricity*. These quantities are designated as e and e'', respectively.

The radius of the graduated circle which passes from point O' through the point O is called the position line of the eccentricity. Let us assume that the instrument is placed so that the position line of the eccentricity crosses the zero mark of the reading microscope J. Let the circle reading be M_0. It is easy to see that the eccentricity has no effect on this reading.

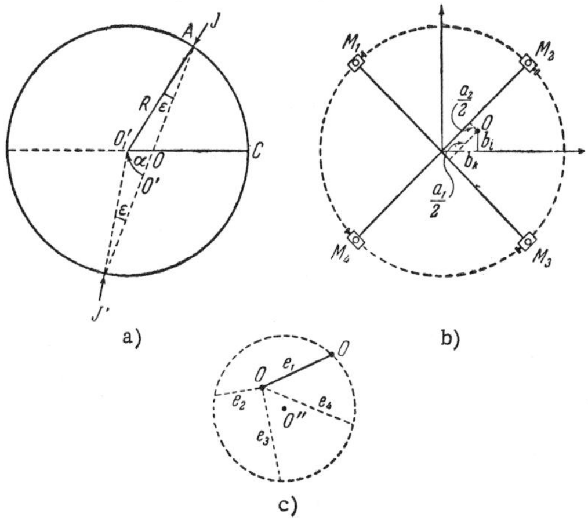

Fig. 37. Eccentricity of the circle.

Let us turn the instrument through some angle α (Fig. 37, a). Then, the position line of the eccentricity is $O_1'C$. The limb of the graduated circle in that particular position is represented by a circle with its center at the point O_1' and radius R. The recorded angle $\alpha' = \angle AO_1'C$ will differ from the actual angle of rotation $\alpha = \angle AOC$ by an error $\varepsilon = \alpha - \alpha'$. It is easy to see that $\varepsilon = \angle OAO_1'$. From the triangle $O_1'AO$ it follows that $R \sin\varepsilon = e \sin(180° - \alpha)$, and consequently $\sin\varepsilon = \frac{e}{R}\sin\alpha$, or, since ε is small, $\varepsilon'' = e''\sin\alpha$.

Consequently, each circle reading of a single microscope M' is altered, due to the circle eccentricity, by the quantity $e''\sin\alpha$; therefore, the correct reading is $M = M' + e''\sin\alpha$.

The periodicity of the eccentricity error can easily be eliminated. This is done by placing a second reading microscope J' at 180° from the first (more precisely, at the other end of the line crossing the center of rotation). Then, the reading of the second microscope will show an error equal in magnitude to the error of the first microscope and opposite in sign. Therefore, the mean reading of the two diagonally opposite microscopes will be free from the error due to eccentricity. This fixes the minimum number of reading microscopes required for astrometric instruments with graduated circles at two.

It is possible to determine the magnitude of the eccentricity and the reading corresponding to the position line of the eccentricity by reading the graduated circle with two microscopes and in several positions of the instrument. For this purpose the instrument is turned several times by a constant angle of, say, 30°. Since the eccentricity is a small quantity, rarely exceeding 10″, it is possible to replace the angle α in the expression $e \sin \alpha$ by the angle α'. Then the difference between the readings M_1' and M_2' of the two diagonally opposite microscopes will be:

$$M_2' - M_1' = [M_2 - e'' \sin(\alpha' + 180°)] - [M_1 - e'' \sin \alpha'] = \\ = M_2 - M_1 + 2e'' \sin \alpha'.$$

Since $M_2 = M_1 + 180° + a$, and $\alpha' = M_1' - M_0$, then,

$$M_2' - M_1' = 180° + 2e'' \sin(M_1' - M_0) + a,$$

where a is a constant quantity which depends on the error in mounting the two microscopes at 180° with respect to each other and is equal to the difference between the readings of the two microscopes when one of them lies along the position line of the eccentricity.

Each setting of the instrument gives us one such conditional equation with the unknowns e, M_0, and a. It may be solved graphically most easily by plotting the quantities $M_2' - M_1'$ against "the reading of the first microscope" on graph paper and drawing a sinusoidal curve through the points. The required parameters are determined from the graph. The constant quantity a, which enters into every reading, does not alter the resulting measured angles; it is regulated by shifting one microscope and must be kept with such limits as will not complicate analysis of the run.

The eccentricity may be investigated by using four microscopes, making it possible to check the centering of the drum with microscopes. In this case, quantities a_1 and a_2, derived for the diameters of two microscopes, specify the distance of the point of intersection of the two diameters from the axis of rotation (Fig. 37, b). By projecting these quantities on the vertical and horizontal axes, it is easy to derive the required drum displacements b_i and b_k in these directions. The angle between the two microscopes is then the average difference between the readings corresponding to the diameters at all circle positions. With these quantities it is possible to center the drum with sufficient accuracy and place the microscopes as required.

The graduated circle is set on a conical surface cut out on the axis and is fastened to it with a washer with screws. By loosening the screws that hold the graduated circle to the instrument axis, it is possible to rotate it on the conical bearing. The axis of this bearing O'' will not, in general, coincide with the center of the circle graduations O'. Hence, if the graduated circle is rotated on the bearing, the linear eccentricity must vary (Fig. 37, c) since the graduation center will describe a circle around the axis of the bearing.

The maximum and minimum values of the linear eccentricity correspond to the circle position at which points O, O', O'' lie along a straight line. If the graduation center lies between the two axes, the eccentricity is a minimum; if outside, the eccentricity is a maximum.

By determining the eccentricity at various positions of the graduated circle on the conical bearing, it is possible to find the position where the eccentricity has the minimum value. It is necessary to note that the procedure for rotating the circle is technically not very simple. Only in very few instruments does the graduated circle rotate easily on the axis with a special key which is subsequently used to lock it in place.

34. Nonperpendicularity of the Plane of the Limb to the Horizontal Axis; the Inclination of the Reading Microscopes

The plane of the limb of the graduated circle must be perpendicular to the axis of rotation of the instrument. The simplest check of this may be made by focusing the reading microscopes with high-power eyepieces. Getting sharp images of the division marks at different circle position (for example, at 30° intervals) and noting the dioptrical reading on the micrometer head of the eyepiece, it is possible to estimate the nonperpendicularity of the limb to the axis of rotation. An exact check is made by using an indicator fixed to the drum and measuring to an accuracy of a few microns the oscillations of the limb surface during the rotation of the instrument.

Usually, in the case of large instruments, the setting of the circles on the instrument axis assures a sufficiently accurate circle position. Let us consider the effect of the obliqueness of the limb, that is, the inclination of the circle, on the microscope reading ΔM_γ.

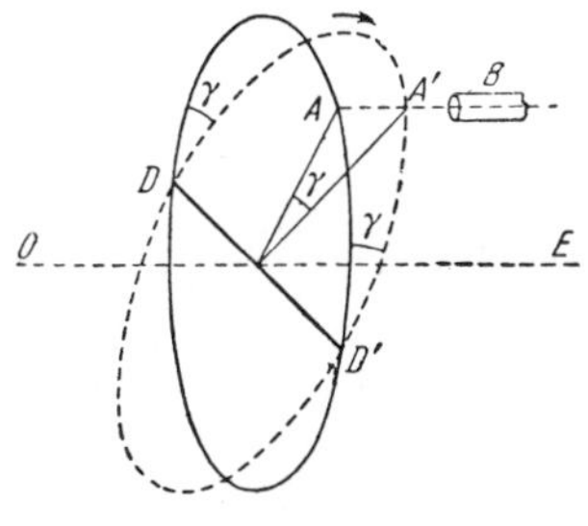

Fig. 38. Nonperpendicularity of the circle to the axis of rotation.

The solid line in Fig. 38 represents the circle when it is perpendicular to the horizontal axis, and the dotted line the circle inclined by an angle γ to it. DD' is the line of intersection of the planes of both circles. Let the index of the reading microscope be set over the point A of the correctly placed circle, and over the point A' of the inclined circle.

The small unknown quantity we seek, ΔM_γ, will be equal to $A'D - AD$. From the spherical triangle ADA' we have

$$\tan AD = \tan A'D \cos \gamma$$

or

$$\tan AD = \tan A'D\left(1 - 2\sin^2\frac{\gamma}{2}\right).$$

Expanding, we obtain:

$$\tan A'D - \tan AD = 2\sin^2\frac{\gamma}{2}\tan A'D,$$

from which

$$\frac{A'D - AD}{\cos A'D \cos AD} = 2\sin^2\frac{\gamma}{2}\tan A'D.$$

As long as angle γ is small and, consequently, the arcs AD and $A'D$ are nearly equal, one can substitute $A'D$ for AD in the trigonometric functions.

By designating, as usual, the circle reading by M, we have: $A'D = M_A - M_D$. If the angle γ is expressed in minutes of arc, then we finally have in units of seconds of arc:

$$\Delta M''_\gamma = \frac{\gamma^2}{229}\sin 2(M_A - M_D).$$

It is easy to calculate that when $\gamma = 2'$, the maximum effect of the circle inclination on the reading made with two microscopes will be only $0''.016$. When the mean reading of four microscopes spaced at 90° intervals with respect to each other is taken, the effect of the obliqueness of the limb to the axis of rotation is completely eliminated as long as it may be represented by a sinusoidal curve with a 180° period. However, another error is connected with the obliqueness of the limb to the axis of rotation; it originates from the inclined position of the reading microscopes.

The line of sight of a correctly placed reading microscope must be parallel to the axis of rotation of the instrument. Consequently, it must be perpendicular to the divided circle if the circle is correctly set on the horizontal axis. However, this is difficult to do and the optical axis of the microscope will make a small angle with the perpendicular to the circle. This angle may be projected on two mutually perpendicular planes which are also perpendicular to the limb. One plane passes through the division mark on which the microscope is trained, and the other is at right angles to it. An inclination in the first plane does not cause any error in the reading (rather, the effect is only in the second order). The angle in the second plane formed by the projected optical axis of the microscope and the perpendicular to the limb is called the *inclination of the reading microscope.* This may cause a noticeable error in the circle reading if the limb is not perpendicular to the horizontal axis or if the limb surface is not a plane. Each of these conditions causes oscillations of the circle when it is rotated, thus varying the distance between the limb and the reading microscope.

Let the principal points of the objective of the reading microscope be represented by a single point C (Fig. 39), and let the angle $\varkappa$ be the

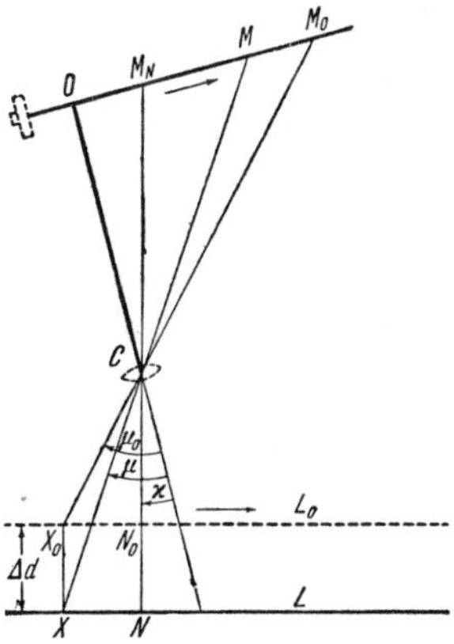

Fig. 39. Inclination of the microscope.

angle between the microscope line of sight and the limb L. Let L_0 be the mean position of the limb with respect to the reading microscope and Δd be the displacement of the limb position from the mean position. The arrows are directed toward increasing limb and micrometer readings. M and M_0 are the readings of some division mark whose positions for the actual and the mean are correspondingly X and X_0. CN_0 is the perpendicular to the limb and M_N is the reading corresponding to the point N_0.

From the triangle XX_0C we have:

$$\frac{\sin(\mu_0-\mu)}{\Delta d}=\frac{\sin(\mu-\varkappa)}{CX_0},$$

and from the triangle X_0CN_0,

$$CX_0=\frac{d}{\cos(\mu_0-\varkappa)}, \text{ where } d=CN_0,$$

from which

$$\sin(\mu_0-\mu)=\frac{\Delta d}{d}\sin(\mu-\varkappa)\cos(\mu_0-\varkappa).$$

The angles $\varkappa$, μ, and μ_0 are small and therefore, up to the second order terms, we have

$$\mu_0-\mu=\frac{\Delta d}{d}(\mu-\varkappa).$$

But since it is possible to assume that the linear displacements of the threads are proportional to the angular displacements,

$$\frac{M_0-M}{\mu_0-\mu}=\frac{M-M_N}{\mu-\varkappa}.$$

We then obtain the relations

$$M_0 - M = \frac{\Delta d}{d}(M - M_N)$$

or

$$\Delta M = \frac{\Delta d}{d} M - \frac{\Delta d}{d} M_N.$$

The first term is just the correction for the change in the run when the limb position changes from the mean position L_0 to the position L. The second term is the correction for the inclination of the microscope and the displacement of the limb with respect to the objective of the reading microscope. As we see, the error due to the inclination operates only in the case when the circle oscillates, that is, when $\Delta d \neq 0$.

To calculate the corrections due to the microscope inclination, it is necessary to determine M_N or $\varkappa$. The quantity Δd is measured by an indicator, as was already mentioned above.

Several methods of investigating the microscope inclination are known. The simplest way to determine it is to change the distance between the limb and the reading microscope by shifting the instrument along the horizontal axis. The instrument is then pointed toward a collimator. The microscope threads are set at the zero point reading and a division mark of the circle is superimposed on them. The instrument is then moved parallel to the horizontal axis a certain amount without disturbing the focus of the reading microscopes. The displacement is measured with an indicator. It is necessary, by using the collimator, to make sure that the instrument does not turn around the horizontal axis. The change of the division reading from one limb position to the other is an indication of inclination of the microscope. By knowing the difference in the readings corresponding to the two positions of the limb as well as the amount of the displacement of the plane of the limb, one can calculate the angle of inclination of the reading microscope.

The autocollimation method for determining the microscope inclination should be mentioned. The inclination of the microscope is determined by the displacement of the light beam reflected from the limb, which changes its direction if the circle oscillates. The effect of the microscope inclination on the circle readings has been investigated in a number of reports.

35. Errors of the Diameters of the Divided Circle and the General Principle for Their Determination

As previously stated, for a number of reasons it is practically impossible to graduate the circle evenly. The accuracy of modern graduating machines for engraving the division marks is close to 3–5 microns. The graduation error, expressed in angular units, depends upon the diameter of the limb; thus, for limb diameters of 60 cm and 100 cm, the error is equal to $2''.0$–$3''.4$ and $1''.2$–$2''.0$, respectively. Hence, it is necessary to determine the error and include it in the reduction of observations.

The term *correction of the division mark* is understood to signify a quantity to be added to the numerical reading of a given division mark in order to obtain the reading actually corresponding to the position of that mark. Usually, by the correction we understand a compensation to be applied with an opposite sign. In practice, it is necessary to know the *diameter correction*, equal to half the sum of the corrections of the division marks located at the ends of the same diameter. This is important since the circle reading is taken with at least two diametrically opposed microscopes.

The errors of the diameter are divided into *systematic and random errors*. If the diameter errors are plotted on a graph, we usually obtain a curve as shown in Fig. 40, where the points represent the errors of individual diameters. The argument varies between the limits of 0° and 180°.

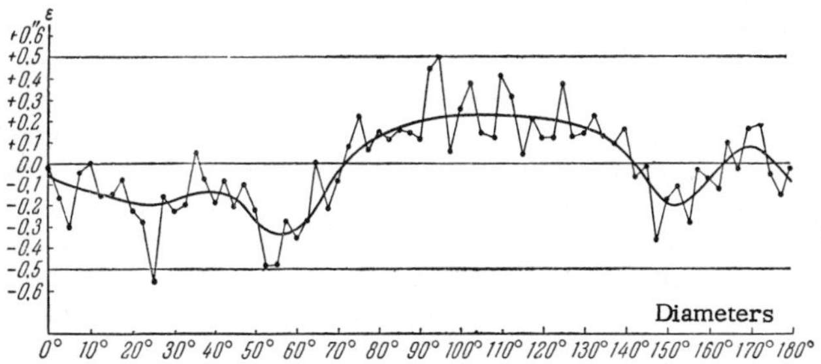

Fig. 40. Errors in the diameters of a graduated circle.

The smooth curve, drawn through the corrections, represents the systematic part of the errors of the diameters. It is the consequence of the smoothly varying errors of the graduating machine and also of the changing external conditions during the process of engraving. The systematic part of the errors of the diameters may be represented by a Fourier series

$$\Delta\varphi = a\cos(2\varphi + A) + b\cos(4\varphi + B) + \ldots,$$

where the argument φ varies between 0° and 360°. There are obviously no odd terms in the expression, since $\Delta\varphi$ is the arithmetic mean of the corrections of the division marks located at the opposite ends of a diameter and separated from each other by 180°. If the readings are taken with four microscopes, then only terms containing the argument in the form 4φ are retained in the expression for $\Delta\varphi$.

The deviations ε of the corrections of individual diameters from the curve of the systematic errors are due to accidental causes existing during the graduation procedure. Such errors are, for example, accidental inaccuracies of the engraver position when it is returned after engraving the division mark. The mean square value of the deviations $\bar{\varepsilon}$ characterizes the graduations in terms of random errors and may be calculated by the formula $\bar{\varepsilon} = \pm\sqrt{\frac{\sum \varepsilon_i^2}{n-1}}$, where ε_i are the deviations of the corrections of individual diameters relative to the smooth error curve and n is the total number of diameter corrections. The quantity ε for modern instruments is approximately $\pm 0''.30$.

The graduation process obviously is not the only source of diameter errors. The changes in the limb of the graduated circle with time (caused by residual random deformations, cleaning of the limb, etc.) alter the diameter corrections. Therefore investigations of the circle must be periodically repeated.

In order to account for the errors in the graduations of the circle, it is necessary to know the corrections for all the diameters. Considering that a limb with 5′ intervals has 4320 division marks and that one with 2′ intervals has 10800 division marks, it is seen that determination of the corrections for all diameters is an extremely laborious task. These corrections were fully determined only for a few graduated circles of astronomical instruments (Greenwich, Paris, Washington). An incomplete study of the limb errors, for example, a determination of diameter corrections at one-degree intervals, enables one to calculate only the systematic part of the corrections corresponding to separate diameters. This may be done graphically by drawing a smooth curve through the observed corrections. Regarding random graduation errors, one is limited only to an estimate of the degree to which they affect circle readings.

The interval between the division mark engraved first and the one engraved last may differ greatly from the average interval between two division marks because of the accumulated engraving errors. When the investigation of the circle is incomplete it is especially important to know at which mark the graduation was begun. In that case, it is particularly necessary to make a special investigation of the section which includes the first and last engraved division marks in order to determine the error due to closing the process of engraving.

With the introduction of automatic graduating machines there appeared the danger of short-period systematic errors, which are sometimes detected. The presence of short-period errors cannot be reconciled with the ordinarily applied methods of incomplete investigation of the limb. Figure 41 shows the periodic errors found in the graduations of the Greenwich meridian circle. The two curves show the values of these errors averaged for two large sections of the limb. The similarity between these curves is evidence of the constant character of these errors with a period of $2^\circ.5$ along the whole circle.

The existing methods for determining the diameter corrections all use the same principle of measurement. Four microscopes are placed at the ends of two diameters which form some angle required in the given method of investigation. Then this angle is measured over different limb sections. The measurement of the angle consists in reading the four microscopes. The readings are not taken immediately after the circle is set at a given position but after a short interval of time, usually one to three minutes. This is necessary to permit the instrument to settle down and maintain the given position during the reading of the four microscopes.

Since it is of basic importance that the angle under measurement remain constant, the best methods are those in which the measurements of the given angle are taken in short sequences. To eliminate possible time-proportional changes in the angle the measurements of each

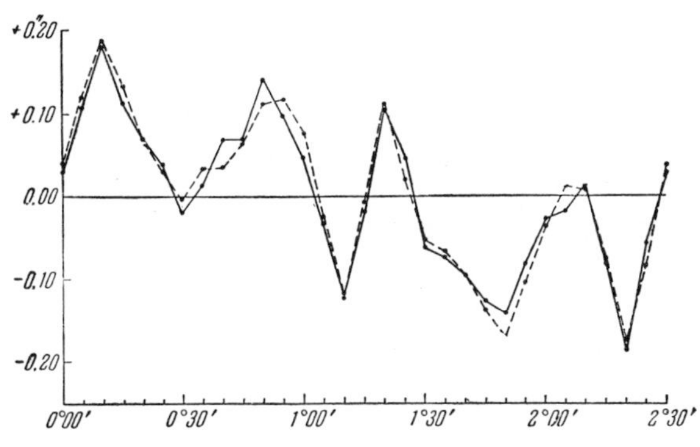

Fig. 41. Short-period errors of the Greenwich circle.

series are repeated in reverse order. The reverse measurements also serve simultaneously as a check on the first readings.

Collimators are used to form angles for measurements in case the astrometric instruments have either two microscopes or four stationary microscopes. The axes of the two collimators are in a plane containing the required angle. The angle is measured with different sections of the limb by rotating the graduated circle along the conic bearing.

The methods for investigating the circles may be divided into those which give an estimate of the accuracy of the circle graduations and those which determine the diameter corrections. These last methods enable one to determine the diameter correction for the whole circle as well as for small sections of the limb.

36. The Heuvelink Method of Estimating the Accuracy of the Circle Graduations

Using the Heuvelink method, it is possible to estimate the accidental errors of the circle graduations and also obtain a general idea of the systematic errors. It is assumed that the systematic errors vary smoothly and, with the readings obtained from two microscopes, can be represented by a Fourier series:

$$\Delta\varphi = a\sin(2\varphi + A) + b\sin(4\varphi + B) + c\sin(6\varphi + C) + \dots$$

The angle α formed by two pairs of microscopes is measured at n circle positions at intervals of $\frac{180^\circ}{n}$. The measurements are taken in direct and reverse sequence in order to minimize the effect of random errors in measurements and also to eliminate changes in α with time.

The average value of the angle α determined at some instrument position is designated by $\bar{\alpha}_i$. Then the true value of the angle α may be represented by the following formula:

$$\alpha = \overline{\alpha}_i + a \sin(2\varphi_i + A) - a \sin(2\varphi_i + 2\alpha + A) + \\ + b \sin(4\varphi_i + B) - b \sin(4\varphi_i + 4\alpha + B) + \\ + c \sin(6\varphi_i + C) - c \sin(6\varphi_i + 6\alpha + C) + \dots$$

We transform this expression by using the formula for the difference of sine terms

$$\alpha = \overline{\alpha}_i - 2a \sin\alpha \cos(2\varphi_i + \alpha + A) - \\ - 2b \sin 2\alpha \cos(4\varphi_i + 2\alpha + B) - \\ - 2c \sin 3\alpha \cos(6\varphi_i + 3\alpha + C) - \dots \tag{14}$$

and expanding the parentheses of the cosines according to the formula for a cosine of the sum

$$\alpha = \overline{\alpha}_i - 2a \sin\alpha \cos(\alpha + A) \cos 2\varphi_i + \\ + 2a \sin\alpha \sin(\alpha + A) \sin 2\varphi_i - \\ - 2b \sin 2\alpha \cos(2\alpha + B) \cos 4\varphi_i + \\ + 2b \sin 2\alpha \sin(2\alpha + B) \sin 4\varphi_i - \\ - 2c \sin 3\alpha \cos(3\alpha + C) \cos 6\varphi_i + \\ + 2c \sin 3\alpha \sin(3\alpha + C) \sin 6\varphi_i - \dots \tag{15}$$

We introduce the notation:

$$\left.\begin{array}{ll} a_1 = 2a \sin\alpha \cos(\alpha + A), & a_2 = -2a \sin\alpha \sin(\alpha + A), \\ b_1 = 2b \sin 2\alpha \cos(2\alpha + B), & b_2 = -2b \sin 2\alpha \sin(2\alpha + B), \\ c_1 = 2c \sin 3\alpha \cos(3\alpha + C), & c_2 = -2c \sin 3\alpha \sin(3\alpha + C) \\ \dots\dots\dots & \dots\dots\dots \end{array}\right\} \tag{16}$$

Upon inserting these quantities into expression (15), we obtain a system of n conditional equations for the unknowns α, a_1, b_1, c_1, ..., a_2, b_2, c_2, ...:

$$\alpha = \overline{\alpha}_i - a_1 \cos 2\varphi_i - a_2 \sin 2\varphi_i - b_1 \cos 4\varphi_i - \\ - b_2 \sin 4\varphi_i - c_1 \cos 6\varphi_i - c_2 \sin 6\varphi_i - \dots$$

We solve these equations by the method of least squares. By multiplying each equation by the coefficient of a particular unknown and summing them up, we obtain the following normal equations:

$$n\alpha = [\overline{\alpha}_i] - [\cos 2\varphi_i]\, a_1 - [\sin 2\varphi_i]\, a_2 - \\ - [\cos 4\varphi_i]\, b_1 - [\sin 4\varphi_i]\, b_2 - [\sin 6\varphi_i]\, c_1 - [\sin 6\varphi_i]\, c_2 - \dots,$$

$$\alpha [\cos 2\varphi_i] = [\overline{\alpha}_i \cos 2\varphi_i] - [\cos 2\varphi_i \cos 2\varphi_i]\, a_1 - \\ - [\cos 2\varphi_i \sin 2\varphi_i]\, a_2 - [\cos 2\varphi_i \cos 4\varphi_i]\, b_1 - \\ - [\cos 2\varphi_i \sin 4\varphi_i]\, b_2 - \dots,$$

$$\alpha [\sin 2\varphi_i] = [\overline{\alpha}_i \sin 2\varphi_i] - [\sin 2\varphi_i \cos 2\varphi_i]\, a_1 - \\ - [\sin 2\varphi_i \sin 2\varphi_i]\, a_2 - [\sin 2\varphi_i \cos 4\varphi_i]\, b_1 - \\ - [\sin 2\varphi_i \sin 4\varphi_i]\, b_2 - \dots \text{ etc.}$$

Since the argument φ is evenly distributed around the circle, the sums of $[\sin 2\varphi_i]$, $[\cos 2\varphi_i]$, $[\sin 4\varphi_i]$, $[\cos 4\varphi_i]$ ... and also the sums of the product functions $[\sin 2\varphi_i \cos 2\varphi_i]$, $[\sin 2\varphi_i \cos 4\varphi_i]$, $[\cos 2\varphi_i \cos 2\varphi_i]$ are equal to zero.

Sums of the form $[\sin 2\varphi_i \sin 2\varphi_i]$, $[\cos 2\varphi_i \cos 2\varphi_i]$, ... are equal to $\frac{n}{2}$.

Taking this into account, it is possible to determine $\alpha = \frac{1}{n}[\bar{\alpha}_i]$. Transforming the normal equation and retaining the terms containing α, we obtain:

$$\alpha[\cos 2\varphi_i] = [\bar{\alpha}_i \cos 2\varphi_i] - \frac{n}{2} a_1,$$

$$\alpha[\cos 4\varphi_i] = [\bar{\alpha}_i \cos 4\varphi_i] - \frac{n}{2} b_1,$$

$$\alpha[\cos 6\varphi_i] = [\bar{\alpha}_i \cos 6\varphi_i] - \frac{n}{2} c_1,$$

. . . .

$$\alpha[\sin 2\varphi_i] = [\bar{\alpha}_i \sin 2\varphi_i] - \frac{n}{2} a_2,$$

$$\alpha[\sin 4\varphi_i] = [\bar{\alpha}_i \sin 4\varphi_i] - \frac{n}{2} b_2,$$

$$\alpha[\sin 6\varphi_i] = [\bar{\alpha}_i \sin 6\varphi_i] - \frac{n}{2} c_2,$$

. . . .

Now it is possible to determine the remaining unknowns

$$a_1 = \frac{2}{n}[(\bar{\alpha}_i - \alpha)\cos 2\varphi_i], \quad a_2 = \frac{2}{n}[(\bar{\alpha}_i - \alpha)\sin 2\varphi_i],$$

$$b_1 = \frac{2}{n}[(\bar{\alpha}_i - \alpha)\cos 4\varphi_i], \quad b_2 = \frac{2}{n}[(\bar{\alpha}_i - \alpha)\sin 4\varphi_i],$$

$$c_1 = \frac{2}{n}[(\bar{\alpha}_i - \alpha)\cos 6\varphi_i], \quad c_2 = \frac{2}{n}[(\bar{\alpha}_i - \alpha)\sin 6\varphi_i]$$

. . . .

As a result of the transformations, the small quantities $(\bar{\alpha}_i - \alpha)$ enter into the derived formulas; this is convenient for calculations.

If a number of nearly equal angles α_j have been measured, then the size of each angle is calculated with its magnitudes $\bar{\alpha}_j$. The unknown quantities a_i, b_i, c_i, ... may be calculated from the complete data consisting of the differences $(\bar{\alpha}_j - \alpha_j)$.

The quantities A, B, C, ... may now be determined from the relations: $\tan(\alpha + A) = -\frac{a_2}{a_1}$, $\tan(2\alpha + B) = -\frac{b_2}{b_1}$, $\tan(3\alpha + C) = -\frac{c_2}{c_1}$ ... because the angle α is known.

The quandrants of the angles $(\alpha + A)$, $(2\alpha + B)$, $(3\alpha + C)$ ... may be found from equations (16) if we consider that a, b, c, ... are positive. The quantities a, b, c, ... are obtained from the relations

$$a^2 = \frac{a_1^2 + a_2^2}{4\sin^2\alpha}, \quad b^2 = \frac{b_1^2 + b_2^2}{4\sin^2 2\alpha}, \quad c^2 = \frac{c_1^2 + c_2^2}{4\sin^2 3\alpha}, \ldots$$

Knowing the quantities α, a, b, c, ... A, B, C, ... and inserting their values into the original equations, it is possible to calculate the residuals

$$\tau = \bar{\alpha}_i - \alpha - \Delta(2\varphi) - \Delta(4\varphi) - \Delta(6\varphi) - \ldots,$$

where $\Delta(2\varphi)$, $\Delta(4\varphi)$, $\Delta(6\varphi)$,... are terms which depend essentially upon the multiples of argument φ. The residuals can be calculated with a different number of terms that depend upon φ. Let us introduce the symbols: $\tau_0 = \bar{\alpha}_i - \alpha$, which is the residual when the periodic terms are neglected; $\tau_1 = \bar{\alpha}_i - \alpha - \Delta(2\varphi)$, which is the residual including the first terms of the periodic series (14); $\tau_2 = \bar{\alpha}_i - \alpha - \Delta(2\varphi) - \Delta(4\varphi)$ is the same quantity with consideration of the first and second terms, etc. From the residuals τ_0, τ_1, τ_2, τ_3 ... it is possible to calculate the root mean square error of one equation

$$\bar{\tau}_0 = \pm\sqrt{\frac{[\tau_0^2]}{n}} = \pm\sqrt{\frac{[(\bar{\alpha}_i - \alpha)^2]}{n}},$$

$$\bar{\tau}_1 = \pm\sqrt{\frac{[\tau_1^2]}{n}} = \pm\sqrt{\frac{[(\bar{\alpha}_i - \alpha - \Delta(2\varphi))^2]}{n}},$$

$$\bar{\tau}_2 = \pm\sqrt{\frac{[(\bar{\alpha}_i - \alpha - \Delta(2\varphi) - \Delta(4\varphi))^2]}{n}}.$$

The quantities $\bar{\tau}_i$ are built up from the errors m, which characterize the measuring process, and the errors s_i of the graduated circle. The quantity m can be determined from repeated measurements of the angle α at one particular circle position:

$$m = \pm\sqrt{\frac{(\alpha_i - \bar{\alpha})^2}{n-1}}.$$

Here $\bar{\alpha}$ is the average value of the angle α as found from these measurements, and n is the number of measurements made. Inasmuch as the quantities $\bar{\tau}_i$ decrease with increasing subscript i, because of a more complete representation of the systematic portion in the division errors, the quantities s_i will also decrease. Beginning with some subscript k, $\bar{\tau}_i$ and, consequently, also s_i will cease diminishing. This indicates that we have reached sufficiently complete representation of the systematic portion of the errors. The subscript k indicates the multiple of φ to which it is necessary to investigate the systematic portion of the error of the circle diameters. From the quantity $\bar{\tau}_k$ one may calculate the root mean square error $\bar{\varepsilon}$ of the circle diameters.

Since the measured angle is distorted by the errors of the two limb diameters, the root mean square error $\bar{\varepsilon}$ of the diameters is related to s_k by the expression $s_k^2 = 2\bar{\varepsilon}^2$ and consequently

$$\bar{\tau}_k^2 = m^2 + 2\bar{\varepsilon}^2 \text{ and } \bar{\varepsilon} = \pm\sqrt{\frac{\bar{\tau}_k^2 - m^2}{2}}.$$

The Heuvelink method does not give the corrections for the real diameters but instead characterizes with sufficient accuracy the random and systematic graduation errors.

37. The Bruhns or Rosette Method

A large number of different methods have been suggested for circle investigations. All of them have some advantages and some disadvantages. However, the method most frequently employed by astronomers is the Bruhns method, also called the rosette method, which gives equal weight to all corrections. The orderliness of the measurements and the simplicity of their processing are characteristic of this method; one disadvantage of it is the large number of readings of each diameter that must be taken—the reason for its high accuracy. There are modifications of the Bruhns method which enable one to reduce the number of measurements and preserve the accuracy of the results. It is possible to calculate the limb characteristics according to the Heuvelink method for measurements made by the Bruhns method.

Besides the rosette method, there is a group of methods based on the method of iteration. These methods—the Perigeau method, for example—are useful for determining the diameter corrections for the whole circle as well as for separate sections of it; however, since the obtained results are not all of the same accuracy, these methods cannot be recommended. According to the method of iteration the corrections for intermediate diameters are derived from the known corrections for some basic diameters. After this, the intermediate diameters are in turn taken as basic diameters and are used for deriving the corrections for another set of intermediate diameters. In this way, by using smaller and smalller limb sections, one obtains corrections for an increasingly large number of diameters. However, the accuracy becomes smaller as the distance between the original standard diameters and the latest observed diameters increases. For this reason, the method is not considered here; those interested should consult M. S. Zverev (see References). The method of iteration is used in cases when other methods are inapplicable due to technical reasons.

We refer to the set of evenly distributed diameters as a rosette. Let us introduce the symbol $R(p, x)$ to describe a rosette consisting of p diameters including a diameter designated by x. Thus, $R(6, 0°)$, signifies the group of diameters at 0°, 30°, 60°, 90°, 120°, 150°.

The investigation of the diameters of the rosette $R(p, x)$ is made with the angles α_i between the diameters of the reading microscopes not exceeding 90° and equal to $\alpha_1 = \frac{180°}{p}$, $\alpha_2 = 2\frac{180°}{p}$, $\alpha_3 = 3\frac{180°}{p}$, etc. These angles are measured with respect to each rosette diameter. Thus, for the case $R(6, 0°)$ the position angles are equal to

$$\alpha_1 = 30°, \quad \alpha_2 = 60°, \quad \alpha_3 = 90°.$$

If we designate the reading of the diameter x by l_x and its correction by (x), then each measurement of angle α_i yields an equation

$$[l_{x+\alpha_i} + (x + \alpha_i)] - [l_x + (x)] = \alpha_i$$

or

$$(x) - (x + \alpha_i) = -\alpha_i + d_{x,\, x+\alpha_i},$$

where $d_{x,\,x+\alpha_i} = l_{x+\alpha_i} - l_x$ is a known quantity. In our particular case, the left hand sides of the equations may be written in the following form:

For angle $\alpha_1 = 30°$ $(0)-(30),\ (30)-(60),\ (60)-(90),$
$(90)-(120),\ (120)-(150),\ (150)-(0);$

For angle $\alpha_2 = 60°$ $(0)-(60),\ (30)-(90),\ (60)-(120),$
$(90)-(150),\ (120)-(0),\ (150)-(30);$

For angle $\alpha_3 = 90°$ $(0)-(90),\ (30)-(120),\ (60)-(150),$
$(90)-(0),\ (120)-(30),\ (150)-(60).$

The measurements of the rosettes for any given angle α_i are made in the direct and reverse orders, with the instrument set at the same readings. If $\alpha = 90°$, the above procedure is not applied since the sequence of measurements taken in one direction already includes two measurements with equal settings.

The corrections can be very easily determined from the above equations. Let us write down, for example, all the equations containing the correction (30):

$$(30)-(60) = -\alpha_1 + d_{30,\,60}, \qquad (30)-(90) = -\alpha_2 + d_{30,\,90},$$
$$(0)-(30) = -\alpha_1 + d_{0,\,30}, \qquad (150)-(30) = -\alpha_2 + d_{150,\,30},$$
$$(30)-(120) = -\alpha_3 + d_{30,\,120},$$
$$(120)-(30) = -\alpha_3 + d_{120,\,30}.$$

If we substract the lower equations from the top ones, we obtain:

$$2(30)-(0)-(60) = d_{30,\,60} - d_{0,\,30} = d_{30}^{30},$$
$$2(30)-(90)-(150) = d_{30,\,90} - d_{150,\,30} = d_{30}^{60},$$
$$(30)-(120) = \frac{1}{2}(d_{30,\,120} - d_{120,\,30}) = \frac{1}{2} d_{30}^{90},$$
$$(30)-(30) = 0,$$

where the subscript designates the diameter whose correction we are seeking, and the superscript designates the angle with which the measurements are carried out. The equations obtained from measurements with an angle of 90° should be taken with the coefficient equal to 1/2. Let us add the last identity. If we combine all the lines, we obtain:

$$6(30) - [(0)+(30)+(60)+(90)+(120)+(150)] =$$
$$= d_{30}^{30} + d_{30}^{60} + \frac{1}{2} d_{30}^{90}.$$

The square brackets contain the sum of the corrections of all the diameters, which we shall in the future designate by S_{30}^{6}; this sum enters the determination of each correction of a diameter and therefore has, as a matter of principle, no meaning. One may impose the condition $\sum(x) = S_x^p = 0$ and derive the correction (30) = $\frac{1}{6}(d_{30}^{30} + d_{30}^{60} + d_{30}^{90})$.

It is not reasonable to apply the unmodified rosette method when a large number of diameters are considered: a great number of angles α_i unnecessarily increases the number of measurements and, similarly, a great number of diameters makes the series of measurements too long; all this also increases the errors in the measurements. Therefore a method developed by Bruhns is used in which the original rosette is broken up into intermediate rosettes.

When the rosettes are combined, the corrections for the rosette with diameters $R(n,x)$ are not derived directly but are obtained by means of the elementary rosettes composed of diameters of the original one. If $n = pqr\ \ldots$, where $p, q, r, \ldots$ are factors of the integer n, then the elementary rosettes corresponding to $R(n, x)$ are $R(p, x), R(q, x), R(r, x), \ldots$. In order to give equal weight to the measurements of all diameters, the rosette $R(p, x)$ is used $\frac{n}{p}$ times, each time starting from a new diameter; similarly, $R(q, x)$ is used $\frac{n}{q}$ times, etc.

Let us clarify this with an example encountered frequently during investigations of graduated diameters of a circle. The rosette is $R(180°, 0°)$ and the diameters to be determined at 0, 1, 2, ... 178, 179°. Since $180 = 9 \times 5 \times 4$, the rosette (180, 0°) will be analyzed by combining the rosettes $R(9, x)$, $R(5, x)$ and $R(4, x)$. It is easy to see that the angles α_i between the microscope diameters are for these rosettes, respectively: 20, 40, 60 and 80°; 36 and 72°; 45 and 90°. It is necessary to use the rosette $R(9, x)$ 180/9 times (that is, 20 times), beginning each series of measurements with one of the diameters at 0, 1, 2 ... 19°. Similarly the rosettes $R(5, x)$ and $R(4, x)$ should be used 36 and 45 times with each series of measurements commencing correspondingly at 0, 1, 2 ... 35° and 0, 1, 2 ... 44°. Using the rosettes $R(9, x)$, $R(5, x)$ and $R(4, x)$, we obtain for the correction of the diameters (x) three equations

$$9(x) - S_x^9 = d_x^{20} + d_x^{40} + d_x^{60} + d_x^{80},$$

$$5(x) - S_x^5 = d_x^{36} + d_x^{72},$$

$$4(x) - S_x^4 = d_x^{45} + \frac{1}{2} d_x^{90};$$

and by combining them

$$18(x) - (S_x^9 + S_x^5 + S_x^4) = F_x; \tag{17}$$

where F_x is the sum of the right hand sides of the equations.

Since different sums $(S_x^9 + S_x^5 + S_x^4)$ enter the expressions of the corrections for different diameters (x), the problem consists in determining the expression in parentheses for each argument using the known quantities F_x.

We form now three sums by combining various values of the quantities F_x:

$$G_x = F_x + F_{x+20} + \cdots + F_{x+160},$$

$$H_x = F_x + F_{x+36} + \cdots + F_{x+144},$$

$$I_x = F_x + F_{x+45} + \cdots + F_{x+135}.$$

If instead of F_x we substitute $18(x) - (S_x^9 + S_x^5 + S_x^4)$, then

$$G_x = (18 - 9) S_x^9 - S_x^{45} - S_x^{36},$$

$$H_x = (18 - 5) S_x^5 - S_x^{20} - S_x^{45},$$

$$I_x = (18 - 4) S_x^4 - S_x^{36} - S_x^{20}.$$

The sums G_x will have 20 values, since $G_0 = G_{20}$, etc.; similarly, H_x and I_x will have 36 and 45 values corresponding respectively to the numbers of rosettes $R(9, x)$, $R(5, x)$ and $R(4, x)$. Hence we obtain:

$$S_x^9 = \frac{1}{9} G_x + \frac{1}{9} S_x^{45} + \frac{1}{9} S_x^{36},$$

$$S_x^5 = \frac{1}{13} H_x + \frac{1}{13} S_x^{20} + \frac{1}{13} S_x^{45},$$

$$S_x^4 = \frac{1}{14} I_x + \frac{1}{14} S_x^{36} + \frac{1}{14} S_x^{20}.$$

In order to determine (x), we insert these expressions in the original equation (17) and get:

$$18(x) - \frac{27}{182} S_x^{20} - \frac{23}{126} S_x^{36} - \frac{22}{117} S_x^{45} =$$
$$= F_x + \frac{1}{9} G_x + \frac{1}{13} H_x + \frac{1}{14} I_x = K_x. \qquad (18)$$

Now we determine S_x^{20}, S_x^{36}, and S_x^{45} using 180 K_x values. For this purpose, we construct the sums L_x, M_x, and N_x.

$$L_x = K_x + K_{x+9} + \ldots + K_{x+171} = \left(18 - \frac{540}{182}\right) S_x^{20} + \frac{9}{180} S_x^{180},$$

$$M_x = K_x + K_{x+5} + \ldots + K_{x+175} = \left(18 - \frac{828}{126}\right) S_x^{36} + \frac{5}{180} S_x^{180},$$

$$N_x = K_x + K_{x+4} + \ldots + K_{x+176} = \left(18 - \frac{990}{117}\right) S_x^{45} + \frac{4}{180} S_x^{180}.$$

There will be, correspondingly, 20, 36, and 45 of these. Hence one may solve for S_x^{20}, S_x^{36}, and S_x^{45} and, eliminating them from equation (18) we obtain:

$$18(x) - \frac{18}{180} S_x^{180} = K_x + \frac{3}{304} L_x + \frac{23}{1440} M_x + \frac{11}{558} N_x.$$

The corrections (x) may be calculated by assuming that $S_x^{180} = 0$. In spite of the fact that the formulas appear to be very complex, the calculations according to the Bruhns method are extremely simple. The theory and application of this method have been discussed in detail by M. S. Zverev (see References).

38. Investigation of Small Sections of the Graduated Circle

In order to know the short-period errors it is necessary to investigate completely several limb sections; then one may also draw a conclusion about the systematic trend of the corrections of the diameters within short intervals of graduation marks. Let us present the method of investigating separate limb sections by applying the principle of the Zurhellen method, which is used to check the uniformity of the graduations of linear scales and, in contrast with the iteration method, gives corrections of equal accuracy. In the Zurhellen method, an auxiliary scale is compared with the scale to be tested, the latter having n

divisions and the auxiliary scale $n-1$ divisions. The scales are laid against each other, with the graduations side by side so that the auxiliary scale may be shifted along the other scale, the latter remaining stationary. The small intervals between the opposing division marks are measured with a microscope micrometer which can also be moved along the scale under examination. The measurements are carried out at several relative scale positions which differ from each other by one interval between the opposing division marks. Instead of the auxiliary scale, a micrometer screw may be used to make settlings on the division marks. Thus, the progressive errors of the screw may be determined as well as the scale corrections.

In the case of a graduated circle one must compare two limb sections with each other, one having n and the other $n-1$ diameters. The relative positions of the sections to be investigated are limited by the minimum angle which the two pairs of microscopes can subtend. This sets the limit not only for the minimum distance but also for the maximum distance between two sections. This method makes it possible to derive corrections for all the diameters of both limb sections as long as one assumes that two of them have no errors. The measurements consist in determining the angular distance between division marks of the two sections with the pairs of microscopes subtending some determined angles. The different angles between the microscope pairs, in the investigation of the circle, correspond to the different relative scale positions.

Let us present this method for the particular case when $n = 5$; it is easy to extend this to any other number of diameters.

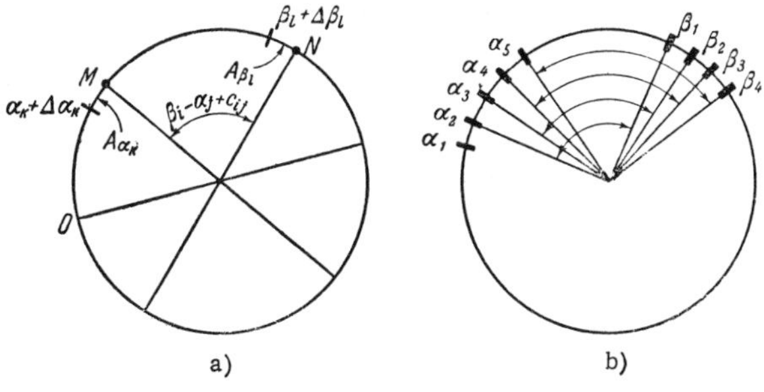

Fig. 42. Investigation of the short-period errors of a circle.

Let the diameters of the first section be identified by α_1, α_2, α_3, α_4, α_5, and of the second by β_1, β_2, β_3, β_4. Let us assume that the angle between the microscopes is equal to $(\beta_i - \alpha_j + c_{ij})$, where c_{ij} is a small quantity. Let A_{α_k} and A_{β_l} be the readings on the two microscopes for diameters α_k and β_l, while $\Delta\alpha_k$ and $\Delta\beta_l$ are the corrections for the diameter identification (Fig. 42). Let us suppose that the numbers increase in going from α to β and that the graduations of the microscope drums increase as the threads move from the lower-numbered to the higher-numbered division marks. Then we can write down the basic equation for the case when the distance between α_k and β_l is measured by the angle $(\beta_i - \alpha_j + c_{ij})$

$$(\beta_l + \Delta\beta_l + A_{\beta_l}) - (\alpha_k + \Delta\alpha_k + A_{\alpha_k}) = \beta_i - \alpha_j + c_{ij}. \tag{19}$$

The structure of this equation can be explained by a diagram (see Fig. 42, a). The diameter M is located at an angular distance $\alpha_k + \Delta\alpha_k + A_{\alpha_k}$ and the diameter N at $\beta_l + \Delta\beta_l + A_{\beta l}$ from the initial diameter. The difference between these angular distances is equal to the angle $\beta_i - \alpha_j + c_{ij}$ between the microscopes.

Since $\beta_i - \alpha_j = \beta_l - \alpha_k$, (19) can be transformed into the form

$$c_{ij} + \Delta\alpha_k - \Delta\beta_l = B_{\alpha_k\beta_l}, \tag{20}$$

where $B_{\alpha_k\beta_l} = A_{\beta_l} - A_{\alpha_k}$ is a quantity determined from measurements. The unknown quantity c_{ij} is the discrepancy in the setting of the angle between the microscopes. It can be eliminated by forming the difference between the two equations in (20), which are derived by using the same angle at two different circle positions:

$$B_{\alpha_k\beta_l} - B_{\alpha_m\beta_n} = \Delta\alpha_k - \Delta\beta_l - \Delta\alpha_m + \Delta\beta_n. \tag{21}$$

We shall successively form the following angles between the microscopes: $(\beta_3 - \alpha_1)$, $(\beta_2 - \alpha_1)$, $(\beta_1 - \alpha_1)$, $(\beta_1 - \alpha_2)$, $(\beta_1 - \alpha_3)$, $(\beta_1 - \alpha_4)$.

With each angle we measure the distance between division marks of the two sections. The circle setting begins with the diameters which define the angles above, and then the circle is rotated until all the possible diameter combinations forming the chosen angle have been obtained. Thus, for example, in the case of the angle $(\beta_1 - \alpha_2)$, the circle must be set into position as shown in Fig. 42, b. For the purpose of symmetry, the measurements can be repeated for every angle in reversed order of instrument positions.

All the quantities obtained by the measurements are represented in the form of Table I:

Table I

		$B_{\alpha_1\beta_1}$	$B_{\alpha_2\beta_1}$	$B_{\alpha_3\beta_1}$	$B_{\alpha_4\beta_1}$
	$B_{\alpha_1\beta_2}$	$B_{\alpha_2\beta_2}$	$B_{\alpha_3\beta_2}$	$B_{\alpha_4\beta_2}$	$B_{\alpha_5\beta_2}$
$B_{\alpha_1\beta_3}$	$B_{\alpha_2\beta_3}$	$B_{\alpha_3\beta_3}$	$B_{\alpha_4\beta_3}$	$B_{\alpha_5\beta_3}$	
$B_{\alpha_2\beta_4}$	$B_{\alpha_3\beta_4}$	$B_{\alpha_4\beta_4}$	$B_{\alpha_5\beta_4}$		
$(\beta_3 - \alpha_1)$	$(\beta_2 - \alpha_1)$	$(\beta_1 - \alpha_1)$	$(\beta_1 - \alpha_2)$	$(\beta_1 - \alpha_3)$	$(\beta_1 - \alpha_4)$

Each column of the table corresponds to measurements taken with an angle whose value is inserted in the last line below each column.

In order to eliminate the unknown quantities c_{ij}, we take the differences between pairs of columns to form Table II:

Table II

$B_{\alpha_5\beta_4} - B_{\alpha_2\beta_1}$	$B_{\alpha_5\beta_3} - B_{\alpha_3\beta_1}$	$B_{\alpha_5\beta_2} - B_{\alpha_4\beta_1}$
$B_{\alpha_1\beta_1} - B_{\alpha_2\beta_2}$	$B_{\alpha_5\beta_4} - B_{\alpha_3\beta_2}$	$B_{\alpha_5\beta_3} - B_{\alpha_4\beta_2}$
$B_{\alpha_1\beta_2} - B_{\alpha_2\beta_3}$	$B_{\alpha_1\beta_1} - B_{\alpha_3\beta_3}$	$B_{\alpha_5\beta_4} - B_{\alpha_4\beta_3}$
$B_{\alpha_1\beta_3} - B_{\alpha_2\beta_4}$	$B_{\alpha_1\beta_2} - B_{\alpha_3\beta_4}$	$B_{\alpha_1\beta_1} - B_{\alpha_4\beta_4}$

If we replace the differences $B_{\alpha_k \beta_l} - B_{\alpha_m \beta_n}$ according to formula (21), we can represent Table II in the form of Table III:

Table III

$\Delta\alpha_5-\Delta\beta_4-\Delta\alpha_2+\Delta\beta_1$	$\Delta\alpha_5-\Delta\beta_3-\Delta\alpha_3+\Delta\beta_1$	$\Delta\alpha_5-\Delta\beta_2-\Delta\alpha_4+\Delta\beta_1$	$+4\,\Delta\beta_1-\sum\Delta\beta-\sum\Delta\alpha+\Delta\alpha_1+4\,\Delta\alpha_5$
$\Delta\alpha_1-\Delta\beta_1-\Delta\alpha_2+\Delta\beta_2$	$\Delta\alpha_5-\Delta\beta_4-\Delta\alpha_3+\Delta\beta_2$	$\Delta\alpha_5-\Delta\beta_3-\Delta\alpha_4+\Delta\beta_2$	$+4\,\Delta\beta_2-\sum\Delta\beta-\sum\Delta\alpha+2\,\Delta\alpha_1+3\,\Delta\alpha_5$
$\Delta\alpha_1-\Delta\beta_2-\Delta\alpha_2+\Delta\beta_3$	$\Delta\alpha_1-\Delta\beta_1-\Delta\alpha_3+\Delta\beta_3$	$\Delta\alpha_5-\Delta\beta_4-\Delta\alpha_4+\Delta\beta_3$	$+4\,\Delta\beta_3-\sum\Delta\beta-\sum\Delta\alpha+3\,\Delta\alpha_1+2\,\Delta\alpha_5$
$\Delta\alpha_1-\Delta\beta_3-\Delta\alpha_2+\Delta\beta_4$	$\Delta\alpha_1-\Delta\beta_2-\Delta\alpha_3+\Delta\beta_4$	$\Delta\alpha_1-\Delta\beta_1-\Delta\alpha_4+\Delta\beta_4$	$+4\,\Delta\beta_4-\sum\Delta\beta-\sum\Delta\alpha+4\,\Delta\alpha_1+\Delta\alpha_5$
$-4\,\Delta\alpha_2+3\,\Delta\alpha_1+\Delta\alpha_5$	$-4\,\Delta\alpha_3+2\,\Delta\alpha_1+2\,\Delta\alpha_5$	$-4\,\Delta\alpha_4+\Delta\alpha_1+3\,\Delta\alpha_5$	

By summing up the columns and setting $\Delta\alpha_1 = \Delta\alpha_5 = 0$, we obtain the diameter corrections multiplied by four, corresponding to the first section α. They are the corrections which evenly distribute the graduations within the interval.

Summing up the lines and making the same assumption, we obtain the quantities $+4\Delta\beta_i - \sum\Delta\beta - \sum\Delta\alpha$. These are the corrections of section β multiplied by four, but actually containing the constant quantity $(-\sum\Delta\beta - \sum\Delta\alpha)$. One should note that the derived corrections distribute the graduations of section β evenly with the scale equal to the scale of the evenly distributed graduations of the first section. This follows from the assumption that c_{ij} is constant inside each series of measurements with a given angle. It follows from the equations for the unknowns $\Delta\alpha_i$ and $\Delta\beta_i$ (see Table III) that they are all determined with equal weights as long as their coefficients are equal.

Comparison of the derived diameter corrections for the case of two distant limb sections (maximum 90°) enables one to judge the invariability of short-period errors along the whole circle.

In order to be able to use the corrections found from the analysis of observations, they should be combined into a general system of corrections obtained for the whole circle, since they were derived on the assumption that the corrections of the end diameters of one of the sections are equal to zero. This is done in the same way: bringing the run of errors of the screw into one system (see Section 41) and applying corrections for these diameters which are derived from the whole circle.

In concluding the description of this method, we note that in its practical execution the microscopes need not be set at all the suggested angles. Actually, measuring several division marks visible in the field of view of the microscope and convenient for measurement is equivalent to setting the microscopes concurrently at several different angles. Thus, if three limb division marks are measured in the field of view of each microscope, corresponding to the diameters α_{i-1}, α_i, α_{i+1} and β_{j-1}, β_j, β_{j+1}, then it is possible, by resetting the circle through three divisions, to derive from the measurements the quantities corresponding to measurements with angles $(\beta_{j+1} - \alpha_{i-1})$, $(\beta_j - \alpha_i)$, and $(\beta_{j-1} - \alpha_{i+1})$. All the systematic errors of the measurements at the edge of the field of view are included in the quantity c_{ij}, and are eliminated when Table II is constructed (on the condition that the measurements are made with the same sections of the screw). This simplifies and speeds up the investigation.

Concluding this presentation of some methods for investigating the graduated circle, we pointed out that further improvement of these methods will entail the use of high-speed computers. It will thus be possible to solve a large number of simple equations interconnecting the diameter corrections. This will permit solution of the problem of determining the corrections for all the diameters of the graduated circle.

CHAPTER VI

The Eyepiece Micrometer

39. The Structure of the Eyepiece Micrometer

The filar micrometer was first used by two Frenchmen, Auzout and Picard, in the year 1666 as an instrument for measuring distances in the focal plane of an astrometric instrument. The first astrometric observations were made by Picard in 1668, installing the micrometer on the quadrant of the Paris Observatory. The meridian instruments set up by Roemer in the period 1672-1681 were also equipped with eyepiece micrometers.

The first eyepiece micrometers had one movable vertical thread and two fixed threads (crossthreads). The structure of the micrometer was improved as time went on. Maskelyne changed the form of the fixed threads in the 18th century by stretching additional vertical threads across the entire field of view. The micrometer for declination measurements was introduced by Bradley. At the end of the 19th century, Repsold proposed a new method for registering right ascensions by adding a contact graduated head to the right ascension screw. Recently, in some instruments, cameras for photographing the micrometer graduated heads have been installed and manual adjustment of the threads in right ascension has been replaced by mechanical adjustment by an electric motor.

The structure of the eyepiece micrometer will be described by using as a model the Repsold registering micrometer, which is most widely used.

The body of the eyepiece micrometer consists of a tube used for attaching it to the instrument tube and for adjustments. In front of it, on the side facing the observer, a lid with guiding accessories is fixed. Collar bearings and bushings of the micrometer screws are attached to the sides of the micrometer tubing. The screw for the right ascension has a graduated head, usually divided into 100 units. Next to this head there is a drum (an auxiliary screwhead for counting the total screw revolutions) which, by means of an epicyclic train of gears, turns by one division for each complete screw revolution. The right ascension screw has a driving gear consisting of a cogwheel and a shaft with two round handles protruding from the micrometer casing. This transmission reduces the number of revolutions; in this way it is possible to make the threads move smoothly and slowly along the right ascension with a relatively fast shaft rotation. Rotation of the right ascension screw is accompanied by displacement of the eyepiece so that the movable threads always remain in the middle of the field of view of the eyepiece. By moving the eyepiece one can observe the star in the best part of the focal plane of the instrument, which would be impossible with a fixed eyepiece having a narrow field of view. The reticle (attached to a frame which can be displaced along the right ascension by a screw) usually has one or two single threads and a double thread. The single threads are used to bisect the star images; the double thread is used for settings on the mercury surface and also for some instrument examinations. The threads are spaced at such intervals that they do not prevent the observer from making a reliable setting and that they are simultaneously visible in the field of view of the eyepiece.

The declination screw is mounted perpendicularly to the right ascension screw. For convenient adjustment in setting the threads in declination, the axis of the drive mechanism is at right angles to the declination screw (this is achieved by the use of bevel gears). The drive handle is located next to the handle for the right ascension screw in order that the observer may easily change from one screw to the other. The net of threads fixed to the frame of the declination screw consists of a single thread for bisecting the star and a double thread for settings on the mercury horizon and the collimator threads. All the stationary threads of the micrometer are fastened on a special fixed frame attached to the back cover of the micrometer. The central part of the field of view is usually free from threads. The fixed vertical threads are symmetrically arranged in groups of five to six on each side of the center, thereby serving to orient the observer's vision in the field of view of the instrument. The horizontal stationary threads form a narrow strip which crosses the center of the field of view of the instrument and defines the place for setting the star image along the vertical direction.

The three thread systems—one fixed and two movable—must be as nearly coplanar as possible (in order to avoid thread parallax), which complicates their construction. The eyepiece has to be focused on the threads so that all the movable threads are clearly visible. Sometimes the threads are attached to only two frames. Then the fixed threads are replaced by additional vertical threads on the movable frame which is operated by the declination screw, and additional horizontal threads on the frame moved by the right ascension screw. These supplementary threads move along their length when the corresponding frames are moved, and hence appear to be fixed.

During daylight observations, a polaroid filter is sometimes placed on the eyepiece in order to decrease the sky background. A reversing prism is used for observations of declinations and right ascensions of stars in the zones near the zenith.

Before we describe the observations with the eyepiece micrometer and their reduction, let us consider the methods of examining the eyepiece micrometer. One usually speaks of micrometer screw errors; however, it is more correct to speak of micrometer errors since they are also attributable to factors other than the screw.

40. Micrometer Errors and the Principle of Their Investigation

Micrometer errors violate the basic principle of micrometers, namely, the proportionality of the linear displacements of the threads to the angle of rotation of the micrometer graduated head. Owing to the rational organization of observations of stars, the micrometer errors do not affect the results of the observations to any great extent. However, they must always be taken into account during the determination of the collimation and during the readings of the mires. We should note that besides the eyepiece micrometer screws, the micrometer screws of the reading microscopes must also be investigated. The micrometer screw errors can be subdivided into periodic and progressive errors.

An uneven value of the screw step at different screw sections, and irregularities in the guides of the movable frame cause the *progressive errors*, i.e., the threads are displaced unevenly during one complete revolution of the screw head at different sections of the screw. The irregularities of the guides contribute to progressive errors only to a certain degree, disturbing the smooth motion of the frame with the thread due to play in the helical threads of the nut. The progressive errors change smoothly along the length of the screw as shown by numerous investigations. This is the consequence of modern techniques of screw manufacture and the use of a nut with a large number of thread turns. The curve of progressive errors frequently has the form of a broken line with a smooth change from one straight line section to another. In high-quality modern screws, the progressive errors do not exceed 0.003-0.006 of a revolution in an interval of 25-30 revolutions.

Periodic errors are errors repeated from one revolution to the next. They depend upon the fractions of the screw revolution and are usually equal for several neighboring turns. This is so because several (about ten) screw thread turns are in simultaneous contact with the threading of the nut fastened to the frame. To determine the constancy of the size of periodic errors for the whole length of the screw, it is necessary to determine them at several sections of the screw; for instance, at the central (working) portion and at the ends.

The causes of periodic errors are rather numerous. The basic cause is an imperfect mounting of the screw in the micrometer frame. The division errors of the graduated head and its eccentric setting on the screw axis also cause periodic micrometer errors. Other causes are noncoincidence of the screw's rotational axis and the thread axis and asymmetry of the screw supports; the latter causes the screw to be displaced periodically along its length.

All these causes bring about an essentially sinusoidal variation of periodic errors with a period of one revolution. Accordingly, we may make the settings on the division marks with two double threads separated by 0.5 or 1.5 revolutions. This eliminates the first term of the series, which is the dominant part of the periodic error (see Section 30). In case of high-quality screws the magnitude of periodic error does not exceed 0.001-0.002 of a revolution.

The general principle for investigating progressive as well as periodic errors of micrometer screws consists in measuring a constant scale interval with different screw sections. The variation of the length of the interval obtained from the measurements can be considered to be the result of screw errors which distort the readings when the micrometer threads are set on the ends of the scale interval. It is very important that the interval length during measurement remain constant. For the investigation of progressive errors we use an interval equal to one screw revolution (or to several, in case of a long screw); for the investigation of periodic errors we need an interval equal to some fraction of one screw revolution. A whole collection of special instruments and procedures has been suggested for micrometer investigations which in one way or another obtain the proper interval and project it on the plane of the micrometer threads. It must be possible to shift the measured interval relative to the screw under investigation, and this is achieved either by shifting the micrometer or the interval itself. The use of these devices requires that the micrometer be detached from the instrument, which is not always convenient and, as a rule, is possible only upon concluding the observations of an entire program.

However, the investigation of the eyepiece micrometer of meridian instruments can be carried out without detaching it from the instrument, and therefore without interrupting a series of observations. In this case the interval to be measured is set up on the frame of the micrometer threads of the meridian collimator. Since the meridian collimators have double micrometers it is possible to investigate the errors of the right ascension screw as well as those of the declination screw. An inadequate stability of the interval during the time of measurement is one of the disadvantages of this method.

41. Determining Progressive Errors and Understanding Their System

From all the methods for determining progressive errors, we present here the most rational method, one frequently applied in practice. In the methods of Lorentzen and Zurhellen, a large number of measurements is necessary to obtain only one result; the method described here gives, for the same number of measurements and with simpler processing, several independent results. Hence it is possible to estimate the precision of the investigation through the external agreement of independent results, which is always preferable to estimating the precision through internal agreement. Application of the Lorentzen and Zurhellen methods is justified only for determining corrections for individual divisions of straight linear scales. In that case individual corrections for each division are found rather than a smooth curve, as in the case of progressive screw errors.

Determination of progressive errors by the method used by geodesists is not desirable for the detailed investigation of micrometer screws, since it approximates the corrections beforehand by a power series which is not always justified.

In order to investigate the progressive errors, the interval approximately equal to one revolution of the screw under investigation is measured by each screw revolution (in case of long screws, the interval is increased to two or three revolutions).

Each measurement of the interval is given by the equation

$$(m_i + \Delta m_i) - (m'_{i-1} + \Delta m_{i-1}) = d, \qquad i = 1, 2, 3, \ldots, n$$

or

$$(m_i - m'_{i-1}) + (\Delta m_i - \Delta m_{i-1}) = d,$$

where m_i and m'_{i-1} are the micrometer readings when it is set on the interval ends, Δm_i are the progressive error corrections, and d the measured interval. In this way, we have a system of n equations resulting from the measurements.

The corrections for the progressive errors must reduce the readings obtained by a screw with irregular pitch to readings of some ideal uniform screw. The size of one revolution of such an ideal screw (scale) and also its position relative to the screw under consideration (zero point) is indeterminate. This is mathematically expressed in the described measuring method by the larger number of unknowns than the number of equations, the excess being equal to two ($n+1$ corrections and d).

A definite solution may be obtained by imposing two conditions on the corrections to be determined; these are referred to as the scale condition and the zero-point condition, their names indicating clearly their nature.

The most convenient scale condition is $\Delta m_0 = \Delta m_n$. Thus one revolution of the ideal screw is assumed to correspond to $\frac{1}{n}$ of n revolutions of the screw under consideration. The size of the measured interval d is then obtained by a straightforward summation of the equations:

$$d = \frac{1}{n} \sum_{k=1}^{n} (m_i - m'_{i-1}).$$

Thus determining the quantity d, we derive the equations for determining the progressive corrections in the following form:

$$\Delta m_i - \Delta m_{i-1} = d - (m_i - m'_{i-1}).$$

We now impose the second condition, the zero-point condition, by assuming for some correction a definitely determined quantity; then we may be able to obtain for the equations a single solution. For the zero-point condition it is usually assumed that $\Delta m_0 = 0$ or $\Delta m_{\frac{n}{2}} = 0$.

However, one should note that the change in the zero-point condition does not affect the shape of the correction curve but only changes all corrections by the same amount. The variation of the condition under which the scale is measured, however, changes the shape of the correction curve; the progressive errors vary according to the ordinate value of some straight line. This straight line passes through a point on the axis of abscissas corresponding to the point of the curve upon which the zero-point condition is imposed. It is clear from this that the linear variation of progressive corrections along the screw does not indicate the presence of the progressive errors but rather that the mean adopted value of one screw revolution is incorrect, that is, the adopted conditions are not correct.

A comparison of progressive error curves has meaning only when the same scale and zero-point conditions are imposed or, stated more briefly, when they belong to the same system. By "system" is meant some concrete scale and zero-point condition.

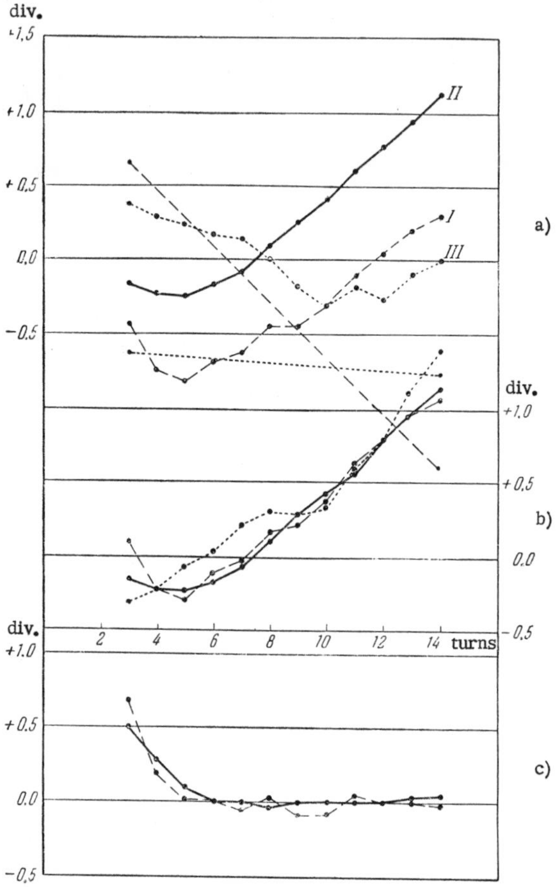

Fig. 43. Progressive errors of a micrometric screw.

In order to illustrate this, we present a comparison of the results of several investigations of a micrometer screw. Figure 43, a presents the graphs of progressive screw errors according to the investigations of three observers: the dashed curve corresponds to the years 1897–98, the solid one to the year 1923, and the dotted one to the year 1924. In comparing these results without bringing them into the same system, the following incorrect deductions were made. The investigations carried out by the first and second observers confirm each other. However, the curve due to the third observer sharply contradicts the previous two investigations in magnitude as well as in sign. The magnitude of the progressive errors is exceptionally large and the quality of the screw will be considered as very poor.

However, before comparing the curves and making deductions, it was necessary to bring them into the same system. In Fig. 43, b the curves are compared after they have been brought into the same system (in this case, the system of the second observer). This translation of the curves is achieved graphically by subtracting from the two curves the ordinates of the corresponding straight lines. As may be seen, the curve of the third observer (dotted curve) is not really a sharp contradiction of the other two investigations. Moreover, it is obvious that the scale specifications were chosen incorrectly. After the correct scale is chosen, the curves take the form shown in Fig. 43, c. Within the interval between the fifth and the fourteenth revolutions the errors are nearly equal to zero, that is, the screw is of good quality within this interval.

42. Determination of Micrometer Periodic Errors; the Kohlschütter Method

The periodic error may be determined by many methods. The most widely used method is the Rydberg method, which superseded the Bessel method used to a great extent in the last century. The Bessel and Rydberg methods of investigation achieve high precision. In the Rydberg method processing of data is simpler and the variation in periodic error is not represented by any particular mathematical form, whereas in the Bessel method the periodic errors are represented by a trigonometric series.

Modern micrometers do not have large periodic errors; consequently it is possible to limit oneself to the first and second order periodic terms which result from imperfect screw mounting of the micrometer. This is the basis of the little known but elegant method of Kohlschütter, which is very simple and also convenient for investigating linear scales. The Kohlschütter method does not require a large number of measurements. In investigating the periodic errors by this method the scale seen in the field of view of the micrometer must be such that one screw revolution corresponds to an integral number of scale divisions. The measurements are achieved by setting the double thread of the micrometer on the division marks of the scale within one revolution of the screw, and with the scale at different positions relative to the screw.

The first series of measurements starts near the zero mark of the graduated head of the micrometer and after one screw revolution it is completed near the zero mark again. In order to eliminate the linear variations in the mounting of the screw the measurements are repeated in the reverse direction. The scale is thereupon shifted so that the first mark to be measured falls at the midpoint of the screw revolution, then the settings are repeated on the marks as before. If one wishes to investigate second-order terms as well, it is necessary to make the series of measurements commencing at 0.25 and 0.75 revolutions.

Let us designate the corrections of the division marks by S_i and the corrections for the periodic screw errors by φ_i. Let the micrometer readings corresponding to settings on the division marks of the scale be m_i. Let the length of the measuring scale, expressed in divisions of the graduated head, be $m_{10} - m_0$. If the value of one scale division corresponds to 0.1 screw revolution the subscript i will change from 0 to 10. Since $\varphi_0 = \varphi_{10}$ due to the nature of periodic errors, then also $S_0 = S_{10}$.

We can now write down the following equations corresponding to each reading:

$$m_0 + \varphi_0 = \frac{0}{10}(m_{10} - m_0) + S_0 + C,$$
$$m_1 + \varphi_1 = \frac{1}{10}(m_{10} - m_0) + S_1 + C,$$
$$m_2 + \varphi_2 = \frac{2}{10}(m_{10} - m_0) + S_2 + C,$$
$$\cdots\cdots\cdots\cdots\cdots\cdots$$

where C is a constant quantity which depends upon the relative positions of the scale and the screw. The correction of one of the scale marks can always be assumed to be equal to zero; for example, $S_0 = 0$ and hence also $S_{10} = 0$. Subtracting from each equation the first equation, we eliminate C, and thus obtain the following nine equations:

$$m_1 - m_0 - \frac{1}{10}(m_{10} - m_0) = S_1 - (\varphi_1 - \varphi_0),$$
$$m_2 - m_0 - \frac{2}{10}(m_{10} - m_0) = S_2 - (\varphi_2 - \varphi_0),$$
$$\cdots\cdots\cdots\cdots\cdots\cdots$$
$$m_9 - m_0 - \frac{9}{10}(m_{10} - m_0) = S_9 - (\varphi_9 - \varphi_0).$$

We would get a similar system of equations from the series of measurements which begin at 0.25, 0.50, and 0.75 screw revolutions.

We designate the left hand terms, which depend on the known measured quantities, by N_1', N_2', ... N_9'. Let us form the average value $\overline{N}_i$ from the four series commencing at 0.00, 0.25, 0.50, and 0.75 revolutions. The summation is carried out in accordance with the subscript of S. Periodic errors of the first and second order are eliminated in forming these average values, as long as they correspond to readings separated from each other by 0.25 and 0.50 revolutions. In this way, the average values $\overline{N}_i$ are equal to the corresponding scale corrections: $\overline{N}_i = S_i$. After eliminating the derived scale corrections, we may write down the following equations for determining the periodic error corrections using all measurements:

$\varphi_1 - \varphi_0 = \overline{N}_1 - N_1'$	$\varphi_{3,5} - \varphi_{2,5} = \overline{N}_1 - N_1''$	$\varphi_6 - \varphi_5 = \overline{N}_1 - N_1'''$	$\varphi_{8,5} - \varphi_{7,5} = \overline{N}_1 - N_1''''$
$\varphi_2 - \varphi_0 = \overline{N}_2 - N_2'$	$\varphi_{4,5} - \varphi_{2,5} = \overline{N}_2 - N_2''$	$\varphi_7 - \varphi_5 = \overline{N}_2 - N_2'''$	$\varphi_{9,5} - \varphi_{7,5} - \overline{N}_2 - N_2''''$
. . . .	. . . .	. . . .	. . . .
$\varphi_9 - \varphi_0 = \overline{N}_9 - N_9'$	$\varphi_{1,5} - \varphi_{2,5} = \overline{N}_9 - N_9''$	$\varphi_4 - \varphi_5 = \overline{N}_9 - N_9'''$	$\varphi_{6,5} - \varphi_{7,5} = \overline{N}_9 - N_9''''$
$\varphi_{10} - \varphi_0 = 0$	$\varphi_{2,5} - \varphi_{2,5} = 0$	$\varphi_5 - \varphi_5 = 0$	$\varphi_{7,5} - \varphi_{7,5} = 0.$

Because these are difference equations, the quantities φ_i are determined to within a constant which has no essential meaning. In order to be able to determine the periodic error corrections uniquely, one must impose within each series the condition $\sum_{10} \varphi_i = 0$. Then, φ_0, $\varphi_{2.5}$, $\varphi_{5.0}$, and $\varphi_{7.5}$ are determined by a simple summation of equations along the columns. For example, for the first series, $\sum_{10} \varphi_i - 10\varphi_0 = \sum_9 (\overline{N}_i - N'_i)$, which gives $\varphi_0 = -\frac{1}{10}\sum_9 (\overline{N}_i - N'_i)$ and the corrections φ_i are obtained from the following expression:

$$\varphi_1 = \overline{N}_1 - N'_1 + \varphi_0; \quad \varphi_2 = \overline{N}_2 - N'_1 + \varphi_0; \ \ldots; \ \varphi_{10} = \varphi_0.$$

Similar calculations are carried out for the remaining three series. Using the corresponding pairs of corrections, we take the mean values.

Two groups of periodic error corrections obtained from different series may differ by a constant quantity because in deriving them the conditions imposed on the error corresponded to different arguments (in one case $\sum \varphi_i = 0$ is chosen for integral tenths, in the other for one and a half tenths). However, as long as the terms higher than the second order are small, the constant may be ignored.

43. Methods of Registering Transit Times of Stars

To determine the right ascensions from observations, it is necessary to obtain, to a high degree of accuracy, the instant at which the star crosses the central thread of the instrument. The older methods of registering the time the star crosses the vertical threads, e.g., the eye-and-ear and the signal-key methods, are almost never applied in astrometry today. The method of registering with a key is sometimes used for observations of stars very near the pole—called "polarissimae" —since, due to the small diurnal motion of these stars, the use of improved methods of recording becomes difficult.

The method of observing with the key is as follows. The observer holds a key similar to a telegraph key in his hand and presses it the moment the vertical thread bisects the star image, thus transmitting a pulse to the chronograph. The transit times are obtained from measurements of the chronograph tape.

The method most frequently used now is the recording method employing a contact micrometer (sometimes also called the registering or impersonal method). As soon as the star appears in the field of view of the instrument, the observer moves it up or down by means of a micrometer adjustment of the telescope tube so that the star image appears to move between two horizontal threads. This position is chosen in order to eliminate the effect of slanting of the vertically moving thread. Afterwards, when the star reaches the movable vertical thread of the micrometer, the observer, using both hands, starts to rotate at a constant rate the driving handle of the micrometer, striving to bisect the star image with the movable thread. While the screw rotates around its axis, pulses are transmitted to the chronograph by means of a toothed drum which makes and breaks a contact (Fig. 44), thus recording those moments during the star's transit which correspond to the moments of make or break of the contact. The arithmetic mean of the times of making or breaking the contacts determined for a certain number of contacts gives the time of transit of the star image through a fictitious central thread. The micrometer reading corresponding to this central thread is called *the middle reading.* Consequently, it should be noted that by the line of sight we understand the straight line joining the second principal point of the objective with the movable micrometer thread set on the middle reading. As may be seen, the recording of transit times by means of a contact micrometer, which demands a certain degree of experience of the observer, to a great extent eliminates personal errors from the observed transit times.

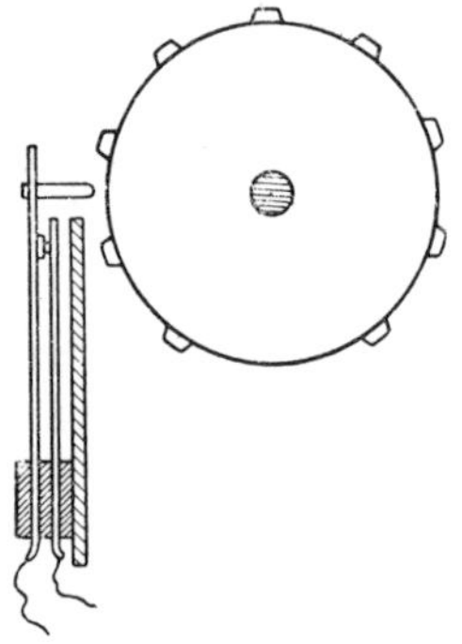

Fig. 44. Contact drum.

Instead of recording by means of contacts, one may photograph the graduated head of the right ascension screw illuminated by flashes from a neon lamp. The flashes are set by the contacts of a clock and hence the readings of the right ascension screw correspond to the star image positions at intervals of one second. This method is convenient because it eliminates the need of recording all but one crossing time on a chronograph in order to identify the number of the second.

Sometimes the so-called mechanical method is used to drive the thread; that is, the thread is moved by a motor. The velocity with which the thread moves depends on the declination of the star, and may be varied either by a rotating cone or by changing the frequency of the current driving a synchronous motor. In the first case, one of the links in the transmission is a rotating cone which has a small wheel pressed against its surface. Depending on whether the wheel is nearer to the vertex or to the base, it will rotate slower or faster and will thus change the speed of the movable thread. The motion of the movable thread is corrected by the observer using the micrometer handle.

Changing the driving speed of the thread by varying the frequency of the current which feeds the motor is the best possible method. In this case the observer need not maintain physical contact with the eyepiece micrometer during the time the thread moves. The observer gives the thread a velocity approximately equal to the speed of the moving star

Fig. 45. Eyepiece micrometer of a modern meridian circle.

and sets it on the image of the star by turning the micrometer handle. Then, he keeps the thread on the middle of the star image, varying the speed of the thread by tuning the frequency generator. One of the defects of this method is the impossibility of constructing a generator with a frequency range including all the velocities of stars with declinations between 0 degrees and the declination of Polaris. This somewhat complicates the reducer gear for transmission of the motor rotation to the micrometer screw, necessitating the use of a gear shift arrangement. The modern eyepiece micrometer is shown in Fig. 45.

44. Photoelectric Method for Recording Stellar Transits

The desirability of impersonal methods of recording transit times led to the introduction of phototubes into astrometry. Astronomer N. N. Pavlov (Pulkovo Observatory) and scientific worker V. E. Brandt (Central Scientific Research Institute for Geodesy, Aerial Photography and Cartography) were occupied for many years in the development of a photoelectric recording method. This method was successfully applied for determination of clock corrections on a small portable reversible transit instrument with a collapsible tube. The practical advantages of this recording method suggest that in the near future it will be used with larger transit instruments and meridian circles for the determination of right ascensions.

The photoelectric method of recording requires some alterations in the instrument. The visual objective lens should be replaced by one whose chromatic curve corresponds to the wavelength sensitivity of the photomultiplier. To be able to fix on the star and make the observations, the instrument must be equipped with a finder, i.e., a small visual telescope mounted parallel to the main one.

The eyepiece micrometer is replaced by a photomultiplier with a lamp and an electrometer, and in the focal plane of the telescope is mounted a metal grid consisting of a series of vertical slits. In its movement past the vertical slits the image of the star causes a variable illumination of the photomultiplier cathode. The current which is generated in the photomultiplier is amplified and the amplified pulses are transmitted to the chronograph. All recorded moments of time correspond to a certain level of signal intensity, thus introducing an element of uncertainty and obliging us to pay particular attention to constancy of signal shape.

It is more practical to place in the focal plane, at an angle of 45° with the optical axis, a grid made on a mirror. The slits are the unsilvered sections of the glass plate. In crossing the slits of the grid light from the star strikes a photomultiplier. In addition to this, the light striking the mirror-like interspaces between the slits is reflected at an angle of 90° to the optical axis and may be directed to a second photomultiplier. The two photomultipliers are connected up to form a bridge and to give a signal at the instant the two light currents are equal. This approximately occurs at the moment when the edge of the slit bisects the star image. In this system the uncertainty of fixing the time is eliminated.

Since the moment of crossing is delayed in this scheme, the time recorded on the chronograph does not correspond to the instant when the light strikes the photomultiplier cathode. Careful investigation of the delay is necessary since it depends on a number of causes: the time, the temperature, etc. For this reason, the moment when the neon lamp in front of the instrument objective is turned on and the flash is registered by the apparatus should be recorded on the chronograph tape.

The main reason for the preference of the photoelectric method is the complete elimination of personal errors; the errors introduced by the new method are smaller and more constant, and are therefore easier to take into account. The role of the observer is reduced to setting the instrument on the star and checking the operation of the apparatus.

Series of observations over several years show that, with respect to accidental errors, the accuracy of photoelectric observations is much higher than that of visual observations. Photoelectric methods do not give large systematic differences, indicating high quality of observations with respect to systematic errors. Among the disadvantages of the photoelectric recording method are the impossibility of observing double stars, planets, and the sun, and the impossibility of its use by day and at twilight. Another disadvantage is the difficulty of aiming the collimators.

45. Determination of Parameters of the Eyepiece Micrometer Required for the Calculation of the Right Ascensions of Stars

The following parameters of the eyepiece micrometer are required for processing of right ascension observations: screw backlash; contact width and the average contact reading; the value of one screw revolution in right ascension; and the distance between the vertical threads.

In reversing the direction of the rotation of the micrometer screw the frame with the threads does not immediately begin to move because of the screw backlash. Therefore two settings made on the same position in the field of view—for example, on the fixed thread—with different directions of screw rotation will give readings differing by an amount equal to the screw backlash. Backlash correction must be applied if observations made at the upper and lower culminations are to be combined. In the presence of backlash all moments of star transit observed at the lower culmination will be shifted relative to those observed at the upper culmination by a quantity $\Delta t_s \sec \delta$, where Δt_s is the backlash expressed in seconds of time. Observations at different instrument positions will differ by this same amount since upon reversing the instrument the direction of screw rotation will also be reversed for observations of similar culminations.

For the same reason, it is necessary to introduce a correction for the contact width. The right ascension screw reading corresponding to the break of a certain contact changes, depending upon the direction of the screw rotation. This occurs because the cog of the contact wheel, having a certain width, touches the contacts at different sides when the wheel rotation is reversed.

The contact width is determined by the following method. A loudspeaker is connected to the micrometer contact outlets in series with a storage battery. Then, at the instant the contact is broken, a sharp click is heard. As the observer slowly rotates the crank driving the screw of the right ascension, at the moment the click occurs he photographs the reading of the right ascension micrometer head. In order to increase the accuracy, measurements are carried out in all screw sections ordinarily used. Following this the observer rotates the screw in the opposite direction and also photographs the readings at the instant the contact breaks. The difference in the readings is the width of a given contact.

Stars are usually observed by turning the screw an integral number of turns; consequently, the average width of the contacts is computed for each turn, although it is possible to determine the width of each contact.

These same measurements permit determination of the average reading of the contacts, that is, the reading corresponding to that position of the movable thread to which the recorded times of individual contacts are brought.

For the processing of observations, it is necessary to know the value of one complete revolution of the eyepiece micrometer screw in angular units; that is, to know to which angle on the celestial sphere a displacement of the micrometer thread caused by one complete screw revolution corresponds. The calibration of one turn of the screw in right ascension is usually made by observing stars with declinations between 60 and 80 degrees. In certain special observations transit times of stars are recorded with as many screw revolutions as is possible. However, more frequently one is limited to using the working screw sections in order not to introduce the progressive errors of the screw and to be able to use the program observations.

If T_1 and T_n are the mean moments of contact for the extreme screw turns, then, from the triangle PAB (Fig. 46), we have: $\sin\ (T_n - T_1) =$

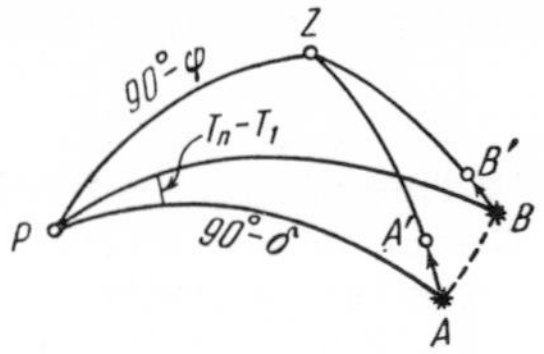

Fig. 46. Determination of the value of one screw turn in right ascension.

$\frac{\sin AB}{\cos\delta}$; the arc $AB = nR_\alpha$, where R_α is the value of one screw turn. By expanding the sines into a series and expressing R_α in seconds of time, we obtain:

$$(T_n - T_1)\cos\delta - \frac{15^2(T_n - T_1)^3}{6\cdot 206\,265^2}\cos\delta = nR_\alpha - \frac{15^2(nR_\alpha)^3}{6\cdot 206\,265^2}$$

or

$$R_\alpha = \frac{(T_n - T_1)}{n}\cos\delta - \frac{15^2}{n\cdot 6\cdot 206\,265^2}\left[(T_n - T_1)^3\cos\delta - (nR_\alpha)^3\right].$$

Assuming that in the last term $nR_\alpha = (T_n - T_1)\cos\delta$, we derive the formula for determining R_α

$$R_\alpha = \frac{(T_n - T_1)}{n}\cos\delta - \frac{15^2}{n\cdot 6\cdot 206\,265^2}(T_n - T_1)^3\cos\delta\sin^2\delta. \tag{22}$$

It is still necessary to calculate the effect of the refraction ρ, which would shift the diurnal parallel towards the zenith and decrease its dimension; hence, the recorded number of screw turns is also decreased and does not correspond to the interval between the observed moments. From the triangles ZAB and $ZA'B'$ we determine respectively the true arc length S and the one distorted by refraction S':

$$S = \Delta A\sin(z' + \rho),$$
$$S' = \Delta A\sin z',$$

where ΔA is the azimuth difference between the two end points of the arc and z is the observed zenith distance. Then

$$\Delta S = S - S' = \Delta A\sin(z' + \rho) - \Delta A\sin z' = $$
$$= 2\Delta A\sin\frac{\rho}{2}\cos\left(z' + \frac{\rho}{2}\right).$$

Since ρ is small, $\Delta S = \Delta A\rho\cos z'$. Taking into account that $\rho = k\tan z'$ and that the average refraction coefficient $k = 57''.4$, we have ΔS in seconds of arc:

$$\Delta S = \frac{1}{206\,265}k\,\Delta A\sin z' = \frac{57.4}{206\,265}S' = \frac{1}{3600}S'.$$

Consequently, after finding the value of one revolution of the screw R_α, we must introduce a correction equal to 1/3600 of R_α. Finally, the value of one screw revolution in seconds of time in right ascension will be equal to:

$$R^s_\alpha = R_\alpha - \frac{1}{3600} R_\alpha,$$

where R_α is calculated by the formula (22).

In order to study the dependence of the value of one screw revolution upon the temperature, it is necessary to calibrate the value of the screw in right ascension at different temperatures. It is sometimes useful to determine the value of one division for each revolution in order to find the progressive errors of the screw.

The distances between the fixed vertical threads are measured in right ascension with the micrometer itself. For this purpose, several settings are made by the movable double thread on each thread separately and readings of the micrometer head are taken for each setting. The average of the readings is taken for each fixed thread and then corrected for the progressive and periodic micrometer screw errors. By knowing the reading of the central thread, the reading corresponding to the meridian and the value of the screw revolution, it is possible to determine from the difference in the readings the distance of each thread from the central thread and from the meridian.

The distances between the vertical threads may also be determined by observing the transits of stars across the threads and recording these moments with a chronograph key. The difference in the transit moments multiplied by cos δ is the distance between the threads in seconds of time.

46. Star Transit across the Meridian of the Instrument

Let us consider the quantities comprising T, the moment of transit of a star across the meridian of the instrument. The basic quantity is the mean transit time $\bar{t}$, obtainable as the arithmetical mean of all the contact times recorded when the star was observed: $\bar{t} = \frac{1}{n} \sum_{i=1}^{n} t_i$.

The individual contact moments t_i are read off the tape of the recording chronograph to an accuracy of $0^s.01$. In order to determine the rate of the recording chronograph with respect to the astronomical clock, a graph is constructed of the values of half-minute contact marks on the chronograph tape, using time as the argument. If there is a regular variation of these values, then it is necessary to draw a smooth curve and take into account the moments $\bar{t}$ deviations of this curve from some constant quantity. If the chronograph does not have its own rate and the clock rate is small, the values of t must remain constant to within the spread of the clock contacts and the chronograph accuracy.

The observations, as a rule, are made on the same screw revolutions in right ascension. Usually, the number of revolutions are chosen such that the number of contacts is between 20 and 30. Observations of circumpolar stars, which move slower, are frequently made with a smaller number of revolutions; during observation of these stars, a supplementary contact wheel is added which has a larger number of contacts.

It is convenient to set the chronograph indicator in advance so that the hour and minute marks will correspond to the sidereal time. This facilitates the designations on the tape, namely, the identification of records for invidual stars and differentiating the contacts of complete revolutions. Because the same number of contacts is used throughout the procedure, it is not necessary to correct $\bar{t}$ for the periodic and progressive errors of the micrometer screw. They are taken into account only when the observations are not carried out on the selected contacts due to a delay in starting the guiding of the star, or if the observation is not made at these contacts for some other reason.

When observations are made at the upper and lower culminations and also at two different instrument positions, the corrections Δt_k for the contact width and Δt_s for the backlash must be introduced.

These corrections are introduced into all observations (taking into account the sign for the upper and lower culminations as well as for both instrument positions), applying one half of the correction in each case.

Finally, the transit time corresponding to the meridian of the instrument may be represented by the sum:

$$T = \bar{t} + \Delta t_\omega + \Delta t_m \sec\delta + \frac{1}{2}\Delta t_k \sec\delta + \\ + \frac{1}{2}\Delta t_s \sec\delta + (c + a)\sec\delta.$$

Here $\bar{t}$ is the average of the moments registered by the separate contacts; Δt_ω is the correction for the rates of the chronograph and of the clock; Δt_m sec δ is the correction for screw error, applied when the usually accepted contact positions are not used; $\frac{1}{2}\Delta t_k$ sec δ is the reduction to the mean reading for contact width; $\frac{1}{2}\Delta t_s$ sec δ is the backlash correction; and $(c + a)$ sec δ is the correction for collimation and diurnal aberration.

The diurnal abberation correction has the form

$$a \sec\delta = -0^s.021 \cos\varphi \sec\delta.$$

In addition, it is necessary to take into account the correction due to irregularities of pivots according to the Mayer formula

$$\Delta i_\zeta \cos z \sec\delta + \Delta k_\zeta \sin z \sec\delta.$$

This correction is sometimes combined with the collimation error by computing a correction table for the irregularities of pivots according to the formula

$$\Delta c_\zeta = \Delta i_\zeta \cos z + \Delta k_\zeta \sin z,$$

where the quantities Δi_ζ and Δk_ζ are the inclination and azimuth corrections of the horizontal axis arising from the pivot irregularities. For the other instrument position corresponding to a certain zenith distance, one takes the corrections with the argument of $360° - \zeta$.

47. Corrections of the Micrometer Reading of Declination for Observations Using Lateral Threads; Calibration of the Declination Screw

In stellar observation the measurement of declination with an eyepiece micrometer is far simpler than recording stellar transits. The thread, movable in declination, is set on the drifting star image, and the micrometer reading is noted. In order to increase the accuracy, several such settings are made at some definite points in the field of view. These points are usually chosen near the fixed vertical threads. Therefore, the expression "observation on the vertical (lateral) threads" is sometimes used.

In order to reduce the separate readings obtained with the lateral threads to the vertical central thread reading set in the meridian, it is necessary to calculate a set of corrections. These are the following: the correction Δm_k for the curvature of the parallel, the correction Δm_J for the slant of the micrometer horizontal thread, and the correction $\Delta m_{J'}$ for the sagging of the same thread. Let us consider each correction separately.

The correction for the curvature of the parallel. The movable horizontal thread is projected on the celestial sphere as a portion of the great circle that crosses the east and west points. Everywhere except at the equator, the star path follows a small circle with the curvature increasing with increasing declination of the star. In this way, the path of the star image in the field of view of the instrument does not coincide with the horizontal thread. Consequently, the declination readings obtained when aligning the horizontal thread on the side threads of the micrometer will differ from the readings obtained by aligning on the central thread, that is, the thread in the plane of the meridian.

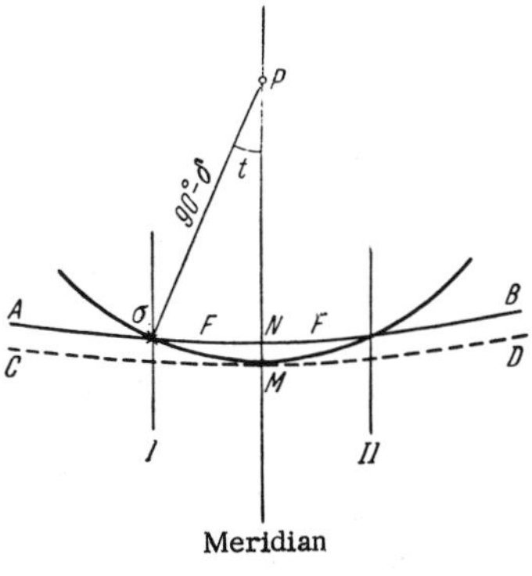

Fig. 47. Curvature of a parallel.

The arc σM (Fig. 47) represents a part of the small circle along which the star moves; AB is the projection of the horizontal movable thread of the micrometer on the celestial sphere, that is, the great circle crossing the east and west points; I and II are projections of the side threads where the settings are made on the star with the horizontal thread. The dashed curve CD represents the position of the projection of the thread when it is set on the star in the meridian. The distance $MN = \Delta m_k$ is the difference in the position of the horizontal thread when it is set on the star in the meridian and on the side threads. Let us determine this quantity.

From the spherical triangle σPN whose angle N equals 90° we obtain:

$$\cos t = \tan \delta \text{ cotan } (\delta + \Delta m_k), \text{ hence } \tan (\delta + \Delta m_k) = \tan \delta \sec t.$$

Let us expand $\tan (\delta + \Delta m_k)$ into a Taylor series and keep only the term with the first derivative since Δm_k is small. Then,

$$\tan \delta + \Delta m_k \sec^2 \delta = \tan \delta \sec t,$$

where Δm_k is expressed in radians, or, transforming, we obtain:

$$\Delta m_k = 2 \tan \delta \sec t \sin^2 \frac{t}{2} \cos^2 \delta.$$

Since t is small, $\sec t \approx 1$, and finally we obtain in units of seconds of arc:

$$\Delta m_k'' = 206\,265'' \cdot \sin^2 \frac{t}{2} \cdot \sin 2\delta.$$

This formula can be transformed if one uses the relation

$$\sin F = \sin t \cos \delta,$$

where F is the distance of the side thread from the meridian. Because angles F and t are small,

$$F = t \cos \delta \text{ and } \Delta m_k = \frac{t^2}{4} \sin 2\delta.$$

From this we obtain in seconds of arc

$$\Delta m_k'' = \frac{15^2}{2} \frac{F^2}{206265} \tan \delta$$

(F is expressed in seconds of time).

The inclination of the movable horizontal thread occurs if the eyepiece micrometer is not mounted exactly. The setting of the eyepiece micrometer is carried out using equatorial stars. By rotating the whole micrometer, we try to keep these stars, drifting along the horizontal thread due to diurnal motion, on the thread. Naturally, some error in the micrometer position still remains.

Let us consider the effect of the inclination of the horizontal thread on the declination reading of the micrometer. Let the solid line in Fig. 48 represent the horizontal thread, inclined at an angle J to the horizon. The dotted line represents the correct position of the horizontal thread as well as the path of an equatorial star in the field of view.

The readings at the middle thread are the same for the inclined and horizontal threads. However, when observing at the side threads I and II the inclined thread must be shifted through the segments Oy and Oz respectively. On the left side, the thread must be displaced one way and on the right side, in the reverse direction; that is, the corrections Δm_J will have opposite signs.

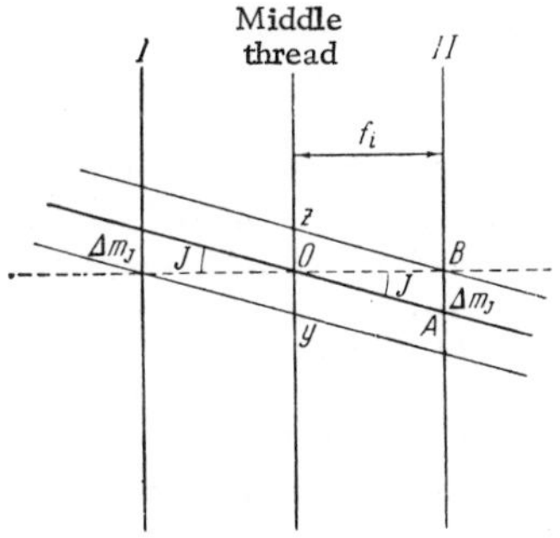

Fig. 48. Inclination of the horizontal thread.

The magnitude of the correction may be derived from the triangle OBA: $\Delta m_J = f_i \tan J$, where f_i is the distance of the side thread from the middle thread. The correction for the thread inclination does not depend upon the star declination. If the observations are made on the same threads, it is possible to omit this correction since the effect of the thread inclination remains constant.

The inclination angle J can be conveniently determined from observations of an equatorial star according to the formula

$$\tan J = \frac{\left(m_\delta^{\mathrm{II}} - m_\delta^{\mathrm{I}}\right) R_\delta}{\left(m_\alpha^{\mathrm{II}} - m_\alpha^{\mathrm{I}}\right) R_\alpha},$$

where m_δ is the declination screw reading when fixing on the star, m_α is the screw reading in right ascension at the points where the declination settings were made, and R_δ and R_α are the values of the micrometer screw revolutions.

Sagging of the horizontal thread is due to the force of gravity and is a very delicate effect, depending upon the quality of stretching of the spider web threads in the micrometer. Poorly tightened threads sag, especially during humid weather. If the thread sags, then this will not affect the readings when the tube is directed towards the zenith since the thread sag will be along the line of sight of the instrument; its effect will increase with a decrease in height of the point under observation and will reach the maximum amount when the instrument position is horizontal. The systematic effect of this error is included in the instrument flexure.

It is possible to detect the thread sag through the difference in the readings resulting by setting the horizontal thread on the dust particles stuck to the vertical threads of the micrometer in the vertical and horizontal positions of the instrument. Since it is difficult to determine the dependence of the thread sag on the zenith distance, it is necessary to stretch the micrometer threads correctly.

In order to determine the value of one screw revolution in declination, observations of stars with well-known declinations are carried out. For this purpose pairs of stars are chosen with differences in declination which can be measured with the screw and with differences in the right ascension not in excess of 10–15 minutes. Such pairs of stars are called "scale pairs."

The instrument is set so that the mean declination of the pair is in the middle of the field of view. Without moving the telescope tube, several settings are made on both stars by the horizontal movable micrometer thread in declination. By correcting each reading for curvature of the parallel in case the stars are observed on different vertical threads, it is possible to form for each star the average reading m. The small difference in right ascension, as a rule, guarantees that the telescope tube will not change its position between the observations of the two stars. As a check one should get the circle readings before observing the first star and after observing the second star.

The value of one screw revolution in declination, in seconds of arc, is equal to

$$R'_\delta = \frac{|\delta_2 - \delta_1|}{|m_2 - m_1|}.$$

In order to obtain the final value of one revolution of the screw it is still necessary to take into account the differential refraction $\Delta\rho$, which is the change in the difference of the declinations arising from different refractional shifts of each star

$$R_\delta = R'_\delta - \frac{\Delta\rho}{|m_2 - m_1|},$$

where $\Delta\rho = k(\tan z_2 \tan z_1)$, k being the refraction coefficient.

The quantity R_δ may also be determined in the laboratory by measuring the shift of the movable horizontal thread by means of a graduated circle. This is done by directing the instrument toward a horizontal collimator. Let the micrometer reading corresponding to the horizontal thread of the collimator be m_1 and the reading of the divided circle be M_1. Let us now turn the instrument by a small angle and take similar readings. Then the angle of rotation of the instrument will be $M_2 - M_1$ according to the divided circle and $m_2 - m_1$ in terms of micrometer revolutions. The value of one screw revolution in declination, in seconds of arc, will be determined from the relation

$$R_\delta = \frac{|M_2 - M_1|}{|m_2 - m_1|}.$$

The screw of the vertical circle may be calibrated by observing the circumpolar stars in elongation. In this method the instant when the star image crosses the horizontal thread is recorded with a chronograph key while the thread is set at different regions of the field of view. The computation is similar to the determination of the value of one screw revolution in right ascension. It is still necessary to take into account the differential refraction and the curvature of the parallel.

48. The Circle Reading

The circle reading M is composed of two parts: the graduated circle reading M' and the correction for the reading of the eyepiece micrometer, ΔM.

The reading of the graduated circle is the average reading $\overline{M}$ of the four microscopes corrected for the average run of the microscopes by $\overline{\Delta M_r}$ and for the graduation errors of the circle by $\overline{\Delta M_\varepsilon}$:

$$M' = \overline{M} + \overline{\Delta M_r} + \overline{\Delta M_\varepsilon}.$$

The average run can be used if the reading microscopes are placed so that the readings of the graduated heads are nearly equal.

The reading of the eyepiece micrometer in declination is $R''_\delta\,(\overline{m}_\delta - m_0)$ where $\overline{m}_\delta$ is the average of the micrometer readings obtained from several settings on the star, m_0 is some rounded-off circle reading introduced to simplify the reduction, and R''_δ is the value of one screw revolution in declination in seconds of arc.

Since, during observations, the star image always lies between two rather closely spaced horizontal threads, the eyepiece micrometer readings in declination are usually, for all stars, within the limits of 0.1 revolution. Hence, it is unnecessary to take into account the periodic and, even less, the progressive screw errors.

It is only necessary to introduce the average correction $\overline{\Delta m_k}$ for the curvature of the parallel and the average thread inclination correction $\overline{\Delta m_J}$. Hence,

$$\Delta M = R''_\delta\,(\overline{m}_\delta - m_0) + \overline{\Delta m_k} + \overline{\Delta m_J}.$$

The quantity ΔM must be added to or subtracted from the quantity M', depending on the direction of the increase of the graduated circle readings with respect to the readings of the eyepiece micrometer in declinations. We may now write the equation for the circle reading:

$$M = M' + \Delta M = \overline{M} + \overline{\Delta M_r} + \overline{\Delta M_\varepsilon} + \\ + R''_\delta\,(\overline{m}_\delta - m_0) + \overline{\Delta m_k} + \overline{\Delta m_J}, \qquad (23)$$

where M' is the corrected reading of the graduated circle and consists of $\overline{M}$—the average of the four microscope readings, $\overline{\Delta M_r}$—the average run correction, and $\overline{\Delta M_\varepsilon}$—the average circle graduation correction; ΔM is the correction of the eyepiece micrometer in declination and consists of $R''_\delta\,(\overline{m}_\delta - m_0)$—the average reading of the eyepiece micrometer in declination, $\overline{\Delta m_k}$—the parallel curvature correction, and $\overline{\Delta m_J}$—the correction for the inclination of the horizontal thread.

In addition, the circle reading must be corrected for flexure (see Section 26).

CHAPTER VII

The Relative or Differential Determinations of Coordinates

49. The Principle of Relative Coordinate Determinations

For the determination of coordinates of a large number of stars the *relative or differential method* is used. In contrast to the absolute method, in which the star coordinates are determined independently from any previously known coordinates, in the differential method the star coordinates are observed relative to the coordinates of some standard reference stars. The coordinates of these reference stars are taken from any first-class fundamental catalog. Thus it is not necessary to define a system of coordinates on the celestial sphere; the system of coordinates is given through standard coordinates obtained from a reference catalog. This accounts for the simplicity of the differential method. The new relative catalog must give the same system of celestial coordinates as the reference star catalog.

The stellar coordinates in any basic catalog contain errors which change more or less smoothly and systematically, depending upon the right ascension and declination of the stars. It is said that a given catalog has a certain "system." Thus, we may say that certain star observations are based on the system FK3, or the relative catalog is in the system of GC, etc.

For relative determinations it is possible to use ordinary meridian instruments; the most convenient of these is the meridian circle (Fig. 49) which determines both coordinates with one star transit. Let us assume that the "ideal" meridian circle is placed exactly in the meridian and that we observe with it two stars: a reference star and the one whose coordinates are to be determined. The difference in the transit times of these stars is equal to the difference in their right ascensions, and consequently, knowing the right ascension of the reference star, we may calculate it for the other star. Similarly the declination of the second star is obtained from the difference in the circle readings of the two stars. This difference is equal to the difference in their declinations, which means that it is necessary only to be able to take into account the errors of the instrument itself as well as of its installation. The constants of the orientation of the instrument as well as their variations are derived from the processing of reference star observation data.

Fig. 49. Greenwich meridian circle.

The observation of relative coordinates is preceded by selection of reference stars, thus compiling the program of observations. The standard coordinates are obtained from the fundamental catalog which is most precise and gives all the necessary data for calculation of the reference star coordinates for any epoch of observation. For modern observations the catalogs are: FK4, N_{30} and GC. Chapter XII is devoted to the description of the fundamental systems.

The reference stars are so chosen that they are evenly distributed in declination and in right ascension. This is necessary, as we shall see further on, in order to be able to tie in easily the new observations with the reference catalog system and for an easy evaluation of changes in the installation of the instrument. The number of reference stars chosen is such that it is possible to design a program for each evening with an optimum ratio of reference stars to the stars of the new catalog, yet without introducing excessive intervals between individual stars on the program. The reference stars usually comprise 30–40% of the total number of stars to be observed during an evening. It is useful to have an extra number of reference stars and combine them in setting up programs for separate evenings. This improves the attachment to the standard system since it minimizes the influence of accidental errors in coordinates of reference stars. The reference stars are evenly distributed with some concentration at the beginning and end of the evening, making it possible to take into account with greater certainty the changes in the instrument position during the evening.

If the list of stars to be newly observed is large, then observations are carried out in declination zones with a width of between several degrees and several tens of degrees, depending on the density of the program. In this case the reference stars are chosen from a wider zone. A better attachment between the new catalog and the fundamental system is achieved if the observations are limited to narrow zones. When observations are made in wide zones, the attachment to a fundamental system is achieved by observing so-called reference sequences, consisting of reference stars only.

50. Relative Determination of Declinations

We shall describe the procedure of observing with a meridian circle. The observer sets the instrument at the approximate zenith distance of the star, some time before its transit, using the finder circle. After the star image appears in the field of view, the observer, by shifting the instrument micrometrically, places the star between the two fixed horizontal threads. The chosen strip is the place within the field of view where all measurements are made; this makes it possible to eliminate certain instrumental errors (see Section 40). For this purpose, the settings of the movable horizontal micrometer thread on the star image are made on all stars in the same portions of the field of view. As a rule, several settings are made, symmetrically distributed with respect to the central thread.

Before the eyepiece micrometer in declination was introduced for actual observing, the star image was set on the fixed horizontal thread by shifting the whole instrument in altitude with micrometric motion; since it was necessary to read off the graduated circle after each setting, the observer had no time to repeat it, and the observations were thus limited to a single setting. The introduction of the eyepiece micrometer permitted an increase in the number of settings for any single star transit in the field of view of the instrument, and thus permitted increasing the accuracy of determination of the declinations.

The instrument must be fixed during observations so that the graduated circle will not shift relative to the drum with the reading microscopes. Consequently, one must use extreme caution when touching the eyepiece micrometer.

After reading the eyepiece micrometer, one should immediately read off the graduated circle using the four microscopes. It is best that both readings be made simultaneously, so that the circle reading will correspond to the instrument position at the instant of observation. This is possible if the circle is read by an assistant or if the photographic method is used. Besides this, the temperature and the air pressure are recorded every hour (in case of observation of stars with $z \gg 50°$, every half hour) in order to be able to calculate the refraction.

When both coordinates are determined together, the star transit is usually observed first, and then the declination eyepiece micrometer is read off. This sequence is used because of the danger of shifting the telescope tube when fixing on the star, thereby changing the circle reading.

The circle reading M is thus obtained from observations of each star. We shall assume that it is corrected not only for all the instrumental errors according to formula (23), but also for refraction. For the determination of refraction it is necessary to calculate the zenith distances using approximate declinations and the latitude. We shall designate the circle readings of the reference stars and the stars under observation by M_j and M_i respectively. Let us assume that the readings increase with increasing declinations. If this is fulfilled in one of the two instrument positions, then to fulfill this condition for the other position as well, it will be necessary to change the readings, taking their 360° supplements instead.

For convenience in taking into account personal and instrumental errors and changes in the orientation of the instrument during an evening of observing, the motion of the "equator point" is introduced. The equator point M_0 is the circle reading obtained when the celestial equator is sighted on. The equator point is related to the circle reading and the declination of the star, as may be easily seen from the following relation: $M_0 = M - \delta$; then, the star declination to be determined is equal to $\delta_i = M_i - M_0$.

Consequently, for the derivation of the declinations it is necessary to be able to determine the equator point. For this purpose reference star observations are used. The apparent declinations of reference stars are calculated from the data of a fundamental catalog and astronomical almanac by application of the rules of spherical astronomy. When the observations are reduced by using the formula $M_0 = M_j - \delta_j$, as many values of M_0 are obtained during the evening as there are reference stars observed. If the reference stars are distributed in a comparatively narrow zone and no changes in position of the instrument have taken place during the course of the evening, the equator points obtained from different stars have a certain scatter arising from accidental errors of observation and random errors in the declinations of reference stars. The declinations to be determined are then calculated with respect to

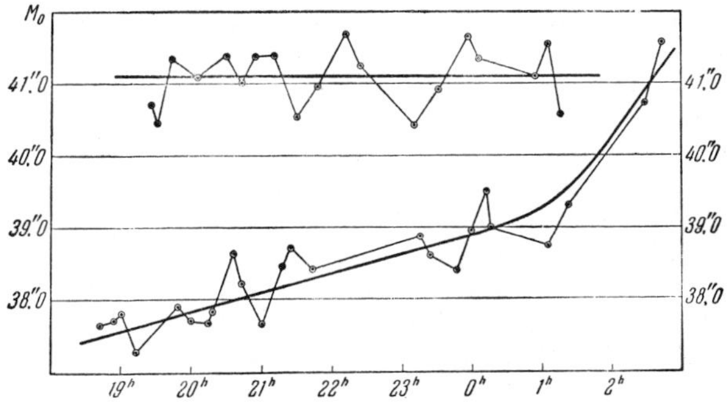

Fig. 50. The drift of the equator points with time.

$\bar{M}_0$, the mean equator point of the evening. Sometimes the equator points show a run with time. In that case, a smooth curve is derived by analytical or graphical means. The chosen values M_0 depend upon the time of observation of the catalog star or upon its right ascension.

The positions of equator points are plotted on Fig. 50 for two observing evenings using the Moscow Repsold meridian circle. On the second graph can be seen the conspicuous drift of the equator points with time and the smooth curve which represents them.

When observations are made within wide zones the scatter of the equator points is caused, in addition to the accidental errors, by systematic errors of observation and in the reference star coordinates.

As shown above, the aim of compiling a catalog of relative observations is the derivation of coordinates for a large number of stars based on an accepted fundamental system. The whole effort must be directed at basing the new catalog as closely as possible on the system of reference stars. This means that the new catalog must have systematic errors identical with those of the fundamental catalog.

51. Determination of the Systematic Difference between the Fundamental and Personal-Instrumental Systems

The difference between the system of the new catalog and the original system may be due to the fact that the astrometric observations made with a given instrument at a certain point on the earth contain systematic errors, personal and instrumental, as well as those caused by local conditions affecting the light path in the atmosphere. Not all these errors can be determined and, consequently, can not be accounted for, due mainly to the complexity of various and sometimes unknown factors which are the basic causes of systematic errors. One should also notice that the organization of observation data and their processing may eliminate the effect of systematic errors of the reference star catalog and thus alter the system of the relative catalog.

In order to find out the course of the change of the system within the zones of the new catalog relative to the fundamental one, and in order to bring the two systems closer together, it is necessary to determine *the systematic difference between the fundamental and the personal-instrumental systems*, that is, "FS-PIS."

The systematic difference FS-PIS can be determined from special observation of reference stars. This method was proposed and first worked out by the German astrometrist Küstner in compiling the Bonn catalog of 1900.

The special series of observations, the so-called reference or Küstner series, are based on reference stars only. The program for each of these series must cover the declination area of all the reference stars and consequently also that of the new catalog. The reference series is observed during the whole course of observations for the new catalog about once a month. The duration of observations of one reference series must be between two to three hours in order to avoid having to take into account the change of the equator point with time. Hence, the number of stars in a reference series could be, for example, 30-40. In compiling a program for a reference series one should endeavor to obtain, if possible, a random sequence of star declinations. Since the reference stars are usually brighter than the program stars, it is convenient to observe the reference series during poor conditions of visibility when observation of fainter stars is inconvenient.

The analysis of the reference series begins by calculating the equator point M_0 from each observation. If, in spite of the compactness of the observations of the reference series, a change with time of the quantity M_0 is noticed, then it is necessary to eliminate this variation from the observations. However, by eliminating this variation, we may distort the dependence of the systematic errors on the declination we are looking for. It is very important to take into account for M_0 all the known instrumental errors, if possible.

Then, within each reference series, the deviations ε_j of individual equator points from their mean value are taken each evening. Sometimes the flexure error of the instrument obtained from an analysis of the change with zenith distance of the quantity ε_j is taken into account. Following this, the deviations ε_j, taken from all the reference series, are grouped according to the declinations they were derived from, ordinarily within five- or ten-degree zones. After calculating the mean values $\bar{\varepsilon}$ for each zone, they are plotted on a graph using the declination as the argument and smoothed graphically in order to reduce the influence of accidental errors. Sometimes, before the smooth curve is drawn, the separate $\bar{\varepsilon}$ values are weighed according to the number of stars used in computing them.

The individual deviations ε_j of the equator point M_0 from its mean value $\overline{M}_0$ for the evening are caused by the following: a) accidental observational errors; b) systematic observational errors; c) accidental errors in star coordinates; d) systematic errors in the system of reference stars. In the course of the analysis, owing to the fact that the observations are averaged out in groups, the accidental errors of observations and of star coordinates are decreased. In smoothing out the curve which relates the variation of $\bar{\varepsilon}$ to the declination, the accidental errors are again decreased. The systematic errors of the observations and of the fundamental catalog enter into the expression $\varepsilon = \overline{M}_0 - M_j + \delta_j$ with opposite signs. The mean of the systematic observational errors in all declinations enter into the quantity $\overline{M}_0 = \overline{M_j - \delta_j}$ with opposite signs. This difference is a constant quantity and enters into each ε_j without changing the progressive variation of ε_j with declination.

The derived smooth curve represents the progressive change with declination of the systematic differences: "fundamental system minus personal-instrumental system." This difference FS-PIS, or γ_i, is often referred to as the "instrument system," since the instrumental errors are the principal part of the "FS-PIS" difference. In case the catalog observations are made by several observers, the difference FS-PIS is derived for each observer separately. Figure 51 shows the difference FS-PIS for declinations and right ascensions of stars of the Fundamental Catalog of Faint Stars (FKSZ) observed on the Repsold meridian circle at Moscow and based on the FK3 system.

In order to study the behavior of the differences FS-PIS during the course of a year, one may analyze the standard series in groups according to the time of the season. As a rule, this is not done in view of the small change of systematic errors and in view of an insufficient

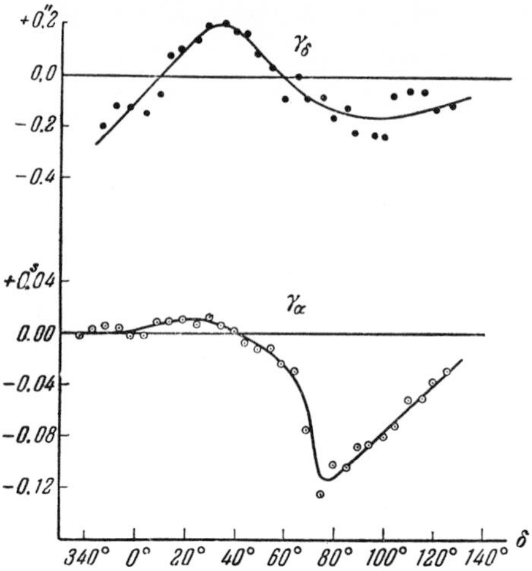

Fig. 51. Progression of the instrument system with declinations and with right ascensions.

number of reference series for obtaining convincing results. The derivation of the systematic difference is sometimes carried out in two approximations. In order to obtain the second approximation to the originally determined differences FS-PIS, the equator points are corrected and then the calculations are repeated. The newly derived values of FS-PIS are considered as corrections to the first approximation.

M. V. Zimmerman proposed a method for determining the FS-PIS difference which, for this purpose, will include (besides the reference series) all observations of the reference stars.

Let us designate by δM_0 the deviations of the equator points from their mean value for the evening. If one ignores the accidental errors, then these quantities differ from the quantities γ_j determined from the reference series by a constant amount $\frac{1}{n}\sum \gamma_l$. The quantities γ_l are found from the declinations of reference stars for a given evening, and hence $\frac{1}{n}\sum \gamma_l$ differ from evening to evening. For reference series, $\frac{1}{n}\sum \gamma_l = 0$ due to the way the quantities γ_l are defined.

For any reference stars observed during an evening, one may write:

$$\delta M_{0j} = \gamma_j - \frac{1}{n}\sum \gamma_l \text{ and } \delta M_{0k} = \gamma_k - \frac{1}{n}\sum \gamma_l.$$

Subtracting, we obtain $\gamma_j - \gamma_k = \delta M_{0j} - \delta M_{0k}$. In this way, it is possible, from a combination of observations of two stars during one evening, to determine not the systematic differences FS-PIS themselves but the differences in their values for two declinations. By combining these differences obtained during all the evenings into groups of five- or ten-degree zones and forming the mean, we obtain the equations of differences for determining the quantities γ_j. Zimmermann proposed an elegant method of solving them.

Let us construct for the quantities $\gamma_j - \gamma_k = \delta M_{0j} - M_{0k}$ the following table:

Declination zones	1	2	...	$n-1$	n
1	$\gamma_1-\gamma_1$	$\gamma_2-\gamma_1$	...	$\gamma_{n-1}-\gamma_1$	$\gamma_n-\gamma_1$
2	$\gamma_1-\gamma_2$	$\gamma_2-\gamma_2$	...	$\gamma_{n-1}-\gamma_2$	$\gamma_n-\gamma_2$
⋮	⋮	⋮	...	⋮	⋮
$n-1$	$\gamma_1-\gamma_{n-1}$	$\gamma_2-\gamma_{n-1}$	...	$\gamma_{n-1}-\gamma_{n-1}$	$\gamma_n-\gamma_{n-1}$
n	$\gamma_1-\gamma_n$	$\gamma_2-\gamma_n$	...	$\gamma_{n-1}-\gamma_n$	$\gamma_n-\gamma_n$
$\sum$	$n\gamma_1-\sum\gamma_i$	$n\gamma_2-\sum\gamma_i$	...	$n\gamma_{n-1}-\sum\gamma_i$	$n\gamma_n-\sum\gamma_i$

The sums of the numbers within each column give the values γ_j for each zone, if the condition $\sum\gamma_i = 0$ is imposed. This method of solving is equivalent to the method of least squares. Its advantages are that all the reference star observations are included in the derivation of FS-PIS and hence increase the accuracy of determining the quantities γ_j. The disadvantage is the amount of work required in compiling the differences $\gamma_j - \gamma_k$.

52. Derivation of Relative Declinations and of the Systematic Difference FS-PIS

The relative declination of a program star is derived from the relation $\delta_i = M_i - M_0$, where the equator point is determined by the formula $M_0 = \overline{(M_j - \delta_j)}$. A change in the position of the equator point with time is excluded. As long as there are systematic differences γ_i = FS-PIS altering the circle readings, these must be corrected by the quantities $(-\gamma_i)$, taken according to the argument of the declination of the observed stars. Since the equator point is obtained from the mean of the obser-variations of n reference stars, the final formula for determining the declinations may be written in the following form:

$$\delta_i = M_i + \gamma_i - M_0 - \frac{1}{n}\sum_{l=1}^{n}\gamma_l. \tag{24}$$

We should remember that the circle readings M_i are corrected for all the known instrumental errors, including the flexure (if it has been investigated), as well as for refraction. This formula points out that it is important to know only the progressive changes of the quantity γ with declination, as long as the constant part is eliminated in the difference

$$\gamma_i - \frac{1}{n}\sum_{l=1}^{n}\gamma_l = \gamma_i - \bar{\gamma}_l.$$

If the observations are made in zones so narrow that the systematic errors of the observations and of the reference catalog may be assumed constant within each zone, it is not necessary to attempt to improve the attachment of the new catalog to the reference system. Actually, if γ = const, then $\gamma_i - \bar{\gamma}_l = 0$.

In case the observations are made within wide zones, where it is impossible to assume that the systematic errors are constant, the corrections for the systematic difference take care of the variation of the systematic errors of observations within each zone; this is true because $\bar{\gamma}_l$ is close to the mean systematic error of the zone, and the difference $\gamma_i - \bar{\gamma}_l$ is the deviation from the systematic error γ_i for a given declination from this quantity. Without these corrections, the system of the new catalog, although close to the fundamental system as far as the mean for each zone is concerned, will still differ from it in details within each zone.

Observations made within wide zones become extreme cases when the observations made each evening embrace the total range of the observed declinations. The mean equator point is then derived from the reference stars having the full range of declinations; hence, the correction $\bar{\gamma}_l$ is equal to zero, owing to the way the quantities γ_i were defined. Computed for each observation according to formula (24), the observed declinations of the selected stars are brought, with the help of an astronomical annual, from the epoch of observation to the beginning of the year of observation, and then, considering precession, to the accepted equinox of the catalog (ordinarily the years 1900.0, 1925.0, 1950.0). In order to determine the final value for the declination, the separate observations of the same star are averaged out.

53. Relative Determination of the Right Ascensions

In order to derive the relative right ascensions from observations it is necessary to determine the transit times of stars. The methods of determining the corrections required due to the instrumental errors were discussed in Chapters IV and VI. The determination of the orientation of the horizontal axis of the instrument and its attachment to the celestial system of coordinates are made by using the known coordinates of reference stars. The observations give us the transit times for the reference stars as well as for the stars of our list, T_j and T_i (see Section 46).

For the reduction of observations, it is convenient to use the Bessel formula

$$\alpha = T + (u + m) + n \tan \delta. \qquad (25)$$

In order to obtain the right ascensions from the observed moments T it is necessary to know $u + m$ and n, which are determined from the observations of reference stars. For the lower culmination the formula is still applicable if one counts the declinations continuously over the pole.

The quantity n is ordinarily obtained from a combination of observations of two reference stars with a large difference of declination. In fact, by subtracting one from the other in expression (25), taken for the two reference stars, we have:

$$(\alpha_j - \alpha_k) - (T_j - T_k) = n\,(\tan\delta_j - \tan\delta_k),$$

from which

$$n = \frac{(\alpha_j - T_j) - (\alpha_k - T_k)}{\tan\delta_j - \tan\delta_k}.$$

Frequently, when observations for the determination of n are made in zones, supplementary reference stars are observed, i.e., the so-called azimuth stars, which are fairly far away from the zone under observation. The observations of the azimuth stars are combined with the observations of reference stars within the zone under observation.

The quantity n is calculated for the evening from several pairs of stars and then the mean value is taken; if many values of n are calculated for one evening, a graph giving the dependence of n on time is constructed. Knowing n, we may calculate for all the reference stars the sum $(u + m)$ using the Bessel formula:

$$(u + m)_j = \alpha_j - T_j - n \tan\delta_j.$$

The unknowns u and m cannot in principle be separated since they have equal coefficients. This simplifies the analysis since it reduces the number of unknowns. The quantity $(u + m)$ is not calculated from observations of azimuth circumpolar stars in view of the large errors in recording their transit times. The obtained values $(u + m)_j$ are either simply averaged out for the evening or are made to fit a smooth curve depending on time, similar to the equator points.

In this way, one either uses the mean values of n and $(u + m)$ for the evening or reads off their values from a graph with time as the argument for any given moment. Now it is easy to calculate the right ascensions for the stars of the program according to the same Bessel formula

$$\alpha_i = T_i + (u + m) + n \tan\delta_i.$$

54. Determination of the Systematic Difference FS–PIS and the Derivation of Relative Right Ascensions

The systematic difference FS–PIS, as in the case of declinations, is determined from reference series. The quantity $(u + m)_j$ is determined in the usual way from observations of each star of a given reference series. The deviations of the separate values $(u + m)$ from their mean for the evening, that is, $\varepsilon_j = \overline{(u+m)} - (u + m)_j$, are grouped according to the declinations for all the series. The values of ε formed in five- or ten-degree zones are arranged in declination and a smooth curve is drawn through these points.

The systematic errors may influence the determined right ascensions through the quantities T and $(u+m)$ (analogous to M and M_0 in the case of declination observations) and also through the quantity n. If one neglects the influence of n, consideration of the systematic difference FS-PIS in right ascensions is not basically different from its consideration in declinations. The correction of right ascensions for the observations during an evening will be equal to $\gamma_i - \frac{1}{s}\sum\gamma_j$, and the calculation of the right ascensions is carried out according to the formula

$$\alpha_i = T_i + \gamma_i + (u+m) - \frac{1}{s}\sum_{j=1}^{s}\gamma_j + n \tan\delta_i,$$

where γ_i are the corrections for the moments T_i taken with the declination of the observed star as the argument. The last but one term is necessary for the correction of $(u+m)$. Indeed, if one assumes that the progressive change with time of the quantity $(u+m)_j$ is taken into account, $(u+m)$ is determined by taking the arithmetical mean of the s values of $(u+m)_j$, i.e., $(u+m) = \overline{\alpha_j} - \overline{T_j} - n \overline{\tan\delta_j}$. The average moment $\overline{T_j}$ must be corrected by the quantity $\overline{\gamma_j} = \frac{1}{s}\sum_{j=1}^{s}\gamma_j$.

If the observations are made within narrow zones, the reference stars used for determination of $(u+m)$ and the stars of the program have declinations so nearly equal that γ = const and, consequently, $\gamma_i = \overline{\gamma_j} = \gamma$. This shows that the system of the new catalog is nearly the same as the system of the reference catalog and does not need any corrections. In case observations are made in wide zones inside which the systematic errors cannot be assumed to be constant, the sum of the corrections $\gamma_i - \overline{\gamma_j}$ will take into account the variation of the systematic errors within the zone, as in the case of declinations.

Let us now consider the effect of the error in the quantity n.

We note that in case observations are made within narrow zones, the errors Δn do not affect the right ascension to be determined regardless of the star declinations from which the quantity n was determined. Actually, the quantity $(u+m)$ includes the error $\Delta n \tan\delta_j$, reduced by the quantity $\Delta n \tan\delta_i$ which distorts the $n \tan\delta$ term of the Bessel formula since for a narrow zone $\tan\delta_i = \tan\delta_j$.

If observations are made in wide zones, the error Δn of the quantity n for a certain evening distorts $(u+m)$ for the observed star; also, the error Δn of the quantity n for the reference series distorts the correction $\gamma_i - \overline{\gamma_j}$ (the error $\Delta n'$ is not equal to n, since, for deriving n for the reference series, stars with all possible declinations were used; but for an ordinary evening only a declination zone is observed). The first effect is analogous to the case of narrow zones, and is expressed in the form $\Delta n\ (\tan\delta_i - \tan\delta_j)$. Since $\gamma_i = \overline{(u+m)} - (u+m)_i$, the effect of $\Delta n'$ on this quantity will be equal to $\Delta n' \tan\delta_i - \Delta n' \tan\overline{\delta} = \Delta n'\ (\tan\delta_i - \tan\overline{\delta})$, and the distortion of $\overline{\gamma_j}$ will be equal to $\Delta n'\ (\tan\overline{\delta} - \tan\delta_j)$. Thus, finally, the remaining effect on α is the member $(\Delta n - \Delta n')(\tan\delta_i - \tan\delta_j)$. The quantity $(\Delta n - \Delta n')$ can be small if in the analysis of all the series, the reference stars and stars of the catalog azimuth stars with approximately the same declinations are used. Moreover, the quantity ($\tan \delta_i - \tan \delta_j$) is smaller than unity in most cases. Thus, although it is in principle impossible to eliminate the effect of the error in the quantity n on the right ascensions to be determined, there is a good reason to assume that this effect is small.

The final apparent right ascensions, obtained from each observation of the stars of the program with the aid of an astronomical almanac, are brought from the date of observation to the beginning of the observation year, and then, by consideration of precession, further reduced to the accepted equinox of the appropriate catalog. The separate values of right ascensions for each star are averaged out.

In concluding this chapter on the determination of relative coordinates, we note that the data from numerous observations of reference stars may be used for decreasing the accidental coordinate errors of these stars.

Let us consider the quantities $\delta M_0 = \overline{M}_0 - M_0$ and $\delta(u + m) = \overline{(u + m)} - (u + m)$ used for the derivation of the systematic differences FS–PIS. The quantities δM_0 and $\delta(u + m)$ contain the difference of the systematic errors of the observations and of the reference catalog, the accidental observation errors, and the individual errors of the reference star coordinates arising from their original observations.

The systematic difference may easily be eliminated by correcting the quantities δM_0 and $\delta(u + m)$ with the derived corrections γ. Averaging observations of a given star over many evenings decreases the accidental observational error so that the individual coordinate corrections of reference stars can be assumed to be equal to

$$\Delta\alpha = \overline{\delta(u+m)+\gamma} - \frac{1}{n}\sum \gamma_i, \quad \Delta\delta = \overline{\delta M_0 + \gamma} - \frac{1}{n}\sum \gamma_j.$$

Calculation of the corrections in the coordinates of reference stars during the reduction of the relative observations is now considered necessary.

CHAPTER VIII

Absolute Determination of Declinations

55. Observations of Stars for Determining the Declinations by the Classical Method. Derivation of the Zenith Distances of Stars

The principle of absolute determinations presupposes that the stellar coordinates may be obtained independently of any previously known coordinates. The relations connecting the unknown declination δ, the zenith distance z, and the latitude φ of the place of observation for a star during its culmination are expressed by three simple formulas: $\delta = \varphi - z$ (the star in upper culmination south of the zenith), $\delta = \varphi + z$ (the star in upper culmination north of the zenith), and $\delta = 180° - \varphi - z$ (the star in lower culmination). Since the latitude φ of the location can be determined without any knowledge of the stellar declinations from observations of zenith distances of circumpolar stars in two culminations (see Section 2), the problem of determining the absolute coordinates reduces basically to measuring the zenith distances.

To observe stars for the purpose of determining declinations, meridian and vertical circles are used. Let us consider how the observations and the determination of zenith distances are made with these instruments.

The principle of determination of zenith distances with a vertical circle was presented in Chapter II. Now we shall describe the process of observation with this instrument.

After setting the vertical circle at the approximate zenith distance, the observer reads off the graduated circle and then reads off the level on the drum or on the frame with the microscopes. After the star appears in the field of view of the instrument, the observer makes several settings on its image with the movable horizontal thread of the eyepiece micrometer. The settings are made on the vertical threads which precede the meridian. After this the instrument is rotated through 180° around the vertical axis and the tube is then turned over the zenith. The settings are then made on the same vertical threads and the readings of the level and the graduated circle are noted. The zenith distance of the star is obtained from these observations with the vertical circle according to formulas of Section 9 with the refraction taken into account.

The method of observing stars with a meridian circle for the purpose of obtaining absolute declinations is not different in any way from the method of relative observations. While the star crosses the field of view of the instrument the observer sets the movable horizontal thread on the star at several determined points in the field of view and reads off the eyepiece micrometer in declination. After this the graduated circle is read.

Throughout the entire evening of observation it is necessary to record the atmospheric temperature and pressure required for the calculations of corrections for refraction. During the absolute determinations of declinations the barometer must be read every hour and the thermometer during observation of every star. The thermometer for measuring the air temperature in the lower layers of the atmosphere must be situated outside the pavilion at the level of the instrument objective. A small telescope is used to read the thermometer. The readings are taken to an accuracy of 0°.02-0°.04 C. The air pressure is measured to an accuracy of 0.05 mm with a mercury barometer, ordinarily situated inside the pavilion. In order to take into account the temperature correction of the mercury in the barometer, the temperature of this mercury is read to an accuracy of 0°.1.

For the determination of declinations when the observations are made with a meridian circle, one must be able to determine the *zenith point*, that is, the circle reading corresponding to the vertical position of the line of sight when it is directed towards the zenith. Instead of the zenith, one may determine the point of the *nadir* of the *horizon*, or of the *pole*, each of which differs from every other one by a constant quantity. Since, in considering the refraction, it is necessary to know the zenith distances, it is preferable to use the zenith point in the analysis. The zenith point may be determined by three methods: a) by means of a mercury surface at the nadir, b) by observing the reflected star images, c) by using the position of the pole obtained from observations of stars nearest the pole—polarissimae. In practice the zenith point is determined most frequently by the first method. The observations of reflected star images have appreciable systematic errors, and the observations of polarissimae have not found widespread application.

The mercury surface used for determination of the zenith point is placed on a stone slab beneath the instrument. In order to set the mercury surface and be able to use it, a trap door is constructed in the floor. The mercury horizon consists of a stand which may be leveled by screws; the mercury has a depth of 1 to 2 mm in a red copper dish. The best dish to use is one with a flat bottom having a diameter equal to the diameter of the objective and with sloping sides making small angles (about 10°-15°) with the bottom. The width of the sides should not be less than half the objective diameter. Such a dish profile assures a fixed position of the mercury perpendicular to the plumb line. Sometimes, in order to decrease the vibration, instead of mercury, certain oils are used which are less likely to vibrate than mercury, or a net is placed in the bottom of the dish to damp the vibration waves appearing on the mercury surface.

If one inserts into the micrometer an autocollimating Gaussian eyepiece and points the instrument tube downward, then, after the light is reflected from the mercury surface, the reflected thread images will be visible in the field of view of the instrument in addition to the micrometer threads. When the movable horizontal thread is superimposed on its reflected image, then the line of sight defined by the thread position is perpendicular to the mercury surface. After the superimposition the declination micrometer and graduation circle are read. From these readings is derived the circle reading corresponding to the nadir or the nadir point. Adding or subtracting 180°, we obtain the zenith point.

The determination of the nadir point is made every hour since it changes during the observing night, mainly due to the temperature changes. The root mean square error of one determination of the nadir point is about $\pm 0''.15$.

The disadvantage of this method is the necessity of rotating the instrument so that the objective points downward. This may cause a systematic difference between these readings and the readings taken at the usual instrument position, that is, with the objective directed upward.

Let us describe other methods for determining zenith distances with the meridian circle.

Fig. 52. Mounting for observing reflected star images.

The zenith point may be eliminated if one observes the star directly as well as its image reflected from the mercury (Fig. 52). This should be done during one and the same transit; thus the graduated circle is read first before the observations, and again after both observations. It is easy to see that, after taking into account the refraction, the circle readings with direct observations and with the reflected images are connected, respectively, to the zenith distances through the following relations: $M_1 = M_z + z$ and $M_2 = M_z + 180° - z$, where M_z is the zenith point. By subtracting the first expression from the second, it is possible to eliminate the zenith point and obtain the zenith distance $z = 90° - \frac{1}{2}(M_2 - M_1)$. This method has not found widespread application because of the observational difficulties arising in practice, especially with stars with small zenith distance. Besides, systematic differences have been observed in declinations obtained with direct and reflected images.

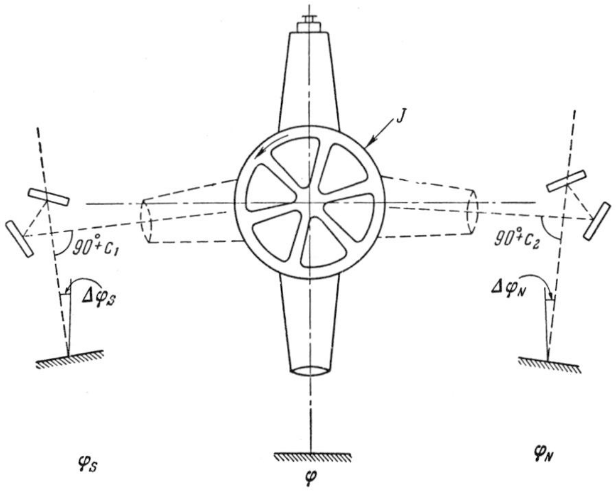

Fig. 53. Determination of the point of the horizon.

The method proposed by M. S. Zverev for the determination of the horizon point deserves to be mentioned. In the meridian of the instrument, at the height of the horizontal axis, reflecting mirrors are placed on both collimator piers with mercury surfaces beneath them (Fig. 53). If the mirrors deflect the light beams exactly by 90°, upon coincidence of the horizontal threads of the eyepiece micrometer and their reflected images the circle reading will correspond to the horizon point. Let $90° + c_1$ and $90° + c_2$ be the angles of deflection of the light by the reflectors, and $\Delta\varphi_S = \varphi - \varphi_S$ and $\Delta\varphi_N = \varphi - \varphi_N$ be the differences between their latitudes and the latitude of the instrument. Then the circle readings M_S and M_N, which correspond to settings on the southern and northern reflectors, and the horizon point M_H, are related as follows:

$$M_H = M_S - c_1 - \Delta\varphi_S \text{ and } M_H = M_N \pm 180° + c_2 - \Delta\varphi_N$$

or

$$M_H = \frac{1}{2}(M_S + M_N \pm 180°) - \frac{1}{2}(c_1 - c_2) - \frac{1}{2}(\Delta\varphi_S + \Delta\varphi_N).$$

The angle of deflection of the light beam caused by the reflectors does not change with the rotation of the reflectors around an axis parallel to the line of intersection of the two mirrors. Therefore, it is possible to transpose the reflectors. We then obtain:

$$M_H = M'_S - c_2 - \Delta\varphi_S \text{ and } M_H = M'_N \pm 180° + c_1 - \Delta\varphi_N$$

or

$$M_H = \frac{1}{2}(M'_S + M'_N \pm 180°) + \frac{1}{2}(c_1 - c_2) - \frac{1}{2}(\Delta\varphi_S + \Delta\varphi_N).$$

Finally we have:

$$M_H = \frac{1}{4}(M_S + M'_S + M_N + M'_N) \pm 90° - \frac{1}{2}(\Delta\varphi_S + \Delta\varphi_N).$$

The last term in this expression is easily calculated. In case the reflectors are situated equidistant from the instrument

$$\Delta\varphi_S = -\Delta\varphi_N \text{ and } M_H = \frac{1}{4}(M_S + M'_S + M_N + M'_N) \pm 90°.$$

The circle reading corresponding to the line of sight directed toward the pole or the polar point may be determined from observations of polarissimae, that is, stars situated very near the celestial pole.

56. Determination of Certain Instrument Constants from Observations of a Polarissima

A star is called a polarissima if it is situated so near the celestial pole that it is possible to observe it in the instrument field of view at almost any time of night.

At the present time, three stars can be observed very near the north celestial pole: BD + 89°1 ($10^m.56$), BD + 89°3 ($9^m.06$), BD + 89°37 ($10^m.06$) at distances from the pole, respectively, 8′, 15′, and 21′. By means of the eyepiece micrometer it is possible to take measurements of no more than 10–15 minutes of arc from the central crossthread. Therefore, the star BD + 89°37 can be observed only near the meridian. The star BD + 89°1 is faint and can be observed only with instruments having an objective with a large diameter. Near the southern celestial pole, there is a circumpolar star CPD-89°38 ($9^m.5$).

The Kiev astronomer V. Fabricius proposed in 1879 to make observations of stars nearest the pole, those natural indicators, in order to derive and check certain instrumental constants. This problem was worked out in theory and practice by L. Courvoisier (Germany) who observed polarissimae for a long time in the first half of the twentieth century. He used these observations for the preparation of a number of catalogs as well as for the derivation of the ephemerides of polarissimae.

During the observations of a polarissima several settings with the horizontal as well as with the vertical thread of the eyepiece micrometer are made and the graduated circle readings in declination and in right ascension are recorded. These observations are made every hour, and are necessarily accompanied by reading the graduated circle and by recording the moments of observation on the chronograph by means of a contact key. The micrometer readings must be corrected for the micrometer screw errors and the circle readings must be corrected for refraction. Since the observations are often made at the edge of the field of view, it is necessary to take into account with great accuracy the inclination of the horizontal thread and the distortion of the object lens.

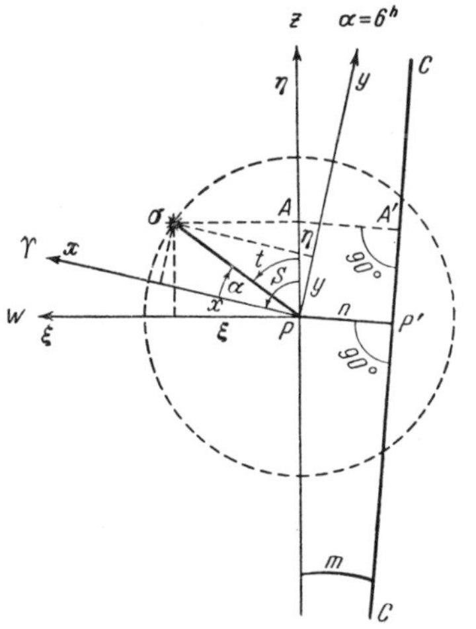

Fig. 54. Illustrating the theory of observation of a polarissima.

The observations of a polarissima permit independent determination of the position of the pole and also the quantity n in the Bessel formula. For the mathematical solution of the problem, in view of the proximity of the polarissima to the pole, it is possible to use two rectangular systems of coordinates with the origin at the celestial pole; one of them, x and y, movable and connected with the celestial sphere and the other one, ξ and η, stationary (Fig. 54). The coordinate x is directed toward the point of vernal equinox and the coordinate y is directed along the declination circle corresponding to $\alpha = 6^h$. The rectangular coordinates x and y of the polarissima are connected to the equatorial coordinates through the relations

$$x'' = 206\,265 \cos\delta \cos\alpha, \quad y'' = 206\,265 \cos\delta \sin\alpha.$$

Since $\delta = 90° - p$, and p is small, then $x'' = p'' \cos\alpha$ and $y'' = p'' \sin\alpha$.

The coordinate η is directed toward the zenith and the coordinate ξ at right angles to it and toward the west. From the triangle $\Delta\sigma PA$, we have:

$$\xi'' = p'' \sin t = p'' \sin(S - \alpha) = x'' \sin S - y'' \cos S,$$
$$\eta'' = p'' \cos t = p'' \cos(S - \alpha) = x'' \cos S + y'' \sin S,$$

where S is the sidereal time and t is the hour angle. The ξ coordinate must be corrected for the diurnal aberration by the quantity $0''.32 \cos\varphi$.

Obviously the circle reading M_σ for a setting on a polarissima differs from the polar point M_P by the quantity η. Thus,

$$M_P = M_\sigma + \eta = M_\sigma + (x'' \cos S + y'' \sin S). \tag{26}$$

The coordinates of a polarissima, x_e and y_e, can be taken for the first approximation from its ephermeris. However, to get absolute values it is necessary to find the corrections Δx_e and Δy_e from observations of the star itself. The determination of the corrections Δx_e and Δy_e is made several times during the evening by micrometric measurement of the coordinates ξ and η. At the same time one must control the stability of the instrument since any shift of the horizontal axis alters the quantities ξ and η. Therefore, it is more appropriate to determine them from the absolute zenith distances of a polarissima as suggested by Courvoisier.

For every observation during one evening it is possible to write down an equation in the form

$$a + \Delta x_e \cos S + \Delta y_e \sin S = \eta_H - \eta_e,$$

where η_H and η_e are the observed and the ephemeris values, and a is some constant depending basically on the systematic errors. The corrections Δx_e and Δy_e can easily be found by the method of least squares. By correcting the ephemeris of the polarissima with these quantities, we can obtain the final position of the polar point from equation (26).

The absolute declination of the observed stars can be calculated from the formula $\delta = |M - M_P - 90°|$, where M is the circle reading with the telescope directed at the star and corrected for refraction.

Observations of a polarissima are useful also for checking the instrument constant n, as was pointed out above. Let CC in Fig. 54 represent the collimation-free line. This line will be removed from the pole P by an amount n, and will be inclined to the meridian by an angle m (see Section 6). Within the accuracy of small terms of higher order, the distance of the polarissima from the collimation-free line is made up of two parts: $\sigma A + AA' = \xi^s + (n^s + \frac{m^s n^s}{13\,751^s})$. On the other hand, from the measurements of the polarissima, this distance is equal to $(m_\alpha - m_c) R^s_\alpha$, where m_α is the micrometer reading in right ascension when a setting is made on the polarissima, m_c is the reading on the collimation-free thread and R^s_α is the value of one screw revolution. Thus we have the following equation for the determination of n:

$$\xi^s + n^s + \frac{m^s n^s}{13\,751^s} = |m_\alpha - m_c|\, R^s_\alpha$$

or, neglecting the term $\frac{m^s n^s}{13\,751^s}$, which is, as a rule, small,

$$n^s = |m_\alpha - m_c|\, R^s_\alpha - \frac{1}{15}(x'' \sin S - y'' \cos S - 0''.32 \cos \varphi).$$

57. Calculation of Preliminary Declinations

The observations with the meridian circle give us the circle reading for the time of the star transit across the meridian M. We assume that this reading has undergone all the corrections, including the eyepiece micrometer reading, according to formula (23). The nadir point $M_N = M_z + 180°$ is determined from the mercury horizon. In order to account for the variation of the zenith point during the night, a graph which shows the dependence of the M_z values on time is constructed. The quantity M_z corresponding to the time of observation of each star is taken from the graph.

After the zenith point is derived, it is possible to determine the observed zenith distance $\zeta = |M - M_z|$. In order to derive the true zenith distance z, it is necessary to take into account the refraction ρ; this is ordinarily done with the Pulkovo refraction tables, which give the refraction according to ζ, the known temperature, and the pressure of the outside air. In this case, the temperature is measured with a psychrometer. The Pulkovo tables of refraction, which have been used for about one hundred years, are in logarithmic form. From the point of view of modern calculating techniques, the tables which give the refraction directly are more convenient. The calculation of refraction is discussed in special literature.

The zenith distances must be corrected for flexure. It is true that for catalog observations one aims to decrease its effect by a suitable arrangement of observations. However, the unavoidable part of flexure still remains and affects the declination system. Consequently the instrument flexure must be carefully studied by laboratory methods and the flexure correction ΔM_ζ taken into account.

When the declinations are determined by the absolute method the preliminary mean latitude φ_0 is usually adopted on the basis of the preceding determinations. The preliminary latitude requires a correction $\Delta\varphi$. Besides the constant correction $\Delta\varphi$, it is necessary to take into account the correction for the variation in latitude arising from the motion of the pole. This phenomenon causes a change of the zenith point on the celestial sphere relative to the equatorial coordinate system; that is, it changes the zenith distances of stars. The effect of the change of latitude can be taken into account with the help of the data given by the International Latitude Service. However, there exists a component of the latitude variation which is caused by purely local conditions. Therefore, it is useful to study the latitude variation by observations with a zenith telescope mounted close to the meridian circle. Observations of stars at two culminations, made with meridian circles for the determination of the mean latitude, are not good for studying this small effect because of insufficient accuracy, but are useful for deriving the constant correction for the mean latitude. The correction of the zenith distance for the variation of the latitude Δz is obtained from observations with the zenith telescope; taken with time as the argument, it is subtracted from the observed zenith distances of stars culminating south of the zenith and added to the stars north of the zenith.

Since the principle of absolute determination must hold, it is necessary to increase the accuracy of the refraction coefficient using the same observations only. The correction of the refraction coefficient Δk is determined together with the correction for the accepted mean latitude $\Delta\varphi$. For this purpose we first calculate the preliminary declinations of the stars which were observed at two culminations according to the formulas

$$\delta_V = \varphi \mp z = \varphi_0 \mp [|M - M_z| + \rho + \Delta M_\zeta \mp \Delta z],$$

$$\delta_N = 180° - \varphi - z = 180° - \varphi_0 - [|M - M_z| + \rho + \Delta M_\zeta + \Delta z].$$

Using the same formulas, the preliminary declinations for all the other stars which were observed are also calculated.

58. Determination of the Corrections of the Mean Latitude and of the Refraction Coefficient

The basic equation for the determination of latitude from observations of zenith distances of one star in two culminations can be written in the following form:

$$90° - \varphi = \frac{1}{2}(z_N \mp z_V), \tag{27}$$

where z_V and z_N are the true zenith distances corresponding to the upper and lower culminations, and the minus sign refers to the case when the upper culmination is south of the zenith.

Let the computed values of refraction be ρ'_N and ρ'_V, while ρ_N and ρ_V are the true values. Since the correction of the refraction coefficient is small, one may adopt the approximate formula $\rho = k \tan z$, representing the change of the refraction with the zenith distance. Then, if we put $x = \frac{\Delta k}{k}$, where k and Δk are the assumed refraction coefficient and its correction, it is possible to write $\rho_N = \rho'_N (1 + x)$ and $\rho_V = \rho'_V (1 + x)$, where x is to be determined. This follows from the equation $(k + \Delta k) \tan z = k \tan z\ (1 + x)$.

We express the zenith distances z in terms of the observes values ζ according to the formulas

$$z_N = \zeta_N + \rho'_N(1 + x) = z'_N + \rho'_N \frac{\Delta k}{k},$$

$$z_V = \zeta_V + \rho'_V(1 + x) = z'_V + \rho'_V \frac{\Delta k}{k},$$

where z'_N and z'_V are the zenith distances at the two culminations determined with the accepted refraction. Substituting the expressions for z_V and z_N in the equation (27) and taking into account that $180° - 2\varphi_0 - (z'_N \pm z'_V) = \delta_N - \delta_V$, we arrive at the basic equation for the determination of the corrections $\Delta\varphi$ and Δk in the form

$$\delta_N - \delta_V = 2\Delta\varphi + (\rho'_N \mp \rho'_V)\frac{\Delta k}{k}. \tag{28}$$

Sometimes, taking into account that $\rho = k \tan z$ while $\delta_N = 180° - \varphi - z$ and $\delta_V = \varphi \mp z$, we transform this equation into the form

$$\delta_N - \delta_V = 2\Delta\varphi + (\tan z_N \pm \tan z_V)\Delta k, \tag{29}$$

where δ_N and δ_V are the preliminary declinations of stars calculated with an accepted latitude and the proper refraction tables.

Equations of the form (28) or (29), set up for each star observed at both culminations, serve as the material for the determination of the unknowns $\Delta\varphi$ and Δk by the method of least squares.

This method for determination of $\Delta\varphi$ and Δk was proposed by Bessel and is considered as the classical method. It was widely used in the 19th and 20th centuries for the derivation of a declination system. However, a study of the quantities $\Delta\varphi$ and Δk obtained from the solution of the equation (28) reveals that they arise from many causes affecting the observed declinations of stars and can be considered only formally to be corrections of the mean latitude and of the refraction coefficient.

On the one hand, the magnitudes of the derived corrections are such that it is impossible to ascribe them to the inaccuracy of the accepted values of the latitude and of the refraction. Thus, the corrections of the mean latitude do not agree with the latitude values obtained from observations with the zenith telescope with which one studies the latitude variations arising from the motion of the pole or other local causes. The refraction coefficient is essentially a physical constant known to a greater degree of accuracy than the scattered set of corrections Δk.

On the other hand, there are some known systematic errors which remain unaccounted for in the analysis and may affect the derived magnitudes $\Delta\varphi$ and Δk. Thus, for example, the unavoidable horizontal flexure $b \sin z$ (see Section 26) affects the left hand side of equation (28) by an amount

$$b \sin z_N + b \sin z_V = 2b \cos \varphi \sin \delta,$$

because $z_N = 180° - \varphi - \delta$, and $z_V = \delta - \varphi$. Since the declinations of the circumpolar stars observed at two culminations are within a sufficiently narrow interval, $\sin \delta$ varies only slightly within this interval, which means that the flexure effect remains practically constant. Consequently, if the horizontal flexure was not eliminated earlier, then its average value will enter into the determined correction $\Delta\varphi$ and will distort it when the equation (28) is solved. A change of the term $2b \cos \varphi \sin \delta$ within the total range of declinations of the circumpolar stars will lead to a fictitious value of Δk.

It is possible to show that correcting the declinations by using $\Delta\varphi$ and Δk, formally calculated according to the Bessel method, in case the flexure has not been eliminated, not only will not improve the declination system but may even make it worse. This effect may be small near the pole but will gradually increase nearer the equator.

Attempts to complicate the equation (28) with supplementary unknowns representing the effect of the flexure error were not successful since the small declination range does not enable us to derive the flexure coefficient to sufficiently high accuracy. An extension of the declination zone by including observations of the sun or planets does not essentially change the weight of the determined corrections.

To sum up, it is proper to realize that there is no satisfactory method for the determination of $\Delta\varphi$ and Δk so long as the flexure has not been eliminated from the observations. One should, consequently, point out that because of unaccounted systematic errors the discrepancies between the declination systems at the equator are of the order of one second of arc in modern catalogs. Therefore the problem of smoothing out a declination system is one of the main problems of fundamental astrometry. Its solution will be possible when new methods are available for eliminating the flexure or when new observational methods which differ basically from the classical method appear.

59. Derivation of Absolute Declinations

The final absolute declinations of observed catalog stars are obtained after the preliminary declinations are corrected using the values $\Delta\varphi$ and x:

$$\delta = \varphi_0 \mp [|M - M_z| + \rho(1+x) + \\ + \Delta M_\zeta \mp \Delta\varphi \mp \Delta\varphi'] \text{ for the upper culminations}$$

$$\delta = 180^\circ - \varphi_0 - [|M - M_z| + \rho(1+x) + \\ + \Delta M_\zeta - \Delta\varphi + \Delta\varphi'] \text{ for the lower culminations}$$

The derived apparent declinations are reduced to the beginning of the year according to the rules of spherical astronomy, and then, by taking precession into account, to the equinox epoch adopted for the catalog. Following this, individual declinations for each star are averaged and the mean epochs of observation for these average declinations are calculated.

If the observations of declinations are made at two instrument positions I and II, then, before combining the declinations obtained at different instrument positions, it is necessary to confirm the absence of systematic differences between them. For this purpose the declination differences $\delta_I - \delta_{II}$ are formed for each star and are averaged over the declination zones (e.g., over five-degree zones). If the differences $\delta_I - \delta_{II}$ vary progressively with declination, the possible causes of this systematic difference must be investigated. If it can be ascertained that the observations at one of the instrument positions are more precise, then the declinations obtained at the other instrument position are reduced to the first one. This is done by adding or subtracting the systematic difference $\delta_I - \delta_{II}$, using as an argument the star's declination. If it is not possible to give preference to observations made at any specific position of the instrument, all declinations are reduced to the mean system of the two positions. If the star was observed an equal number of times at both instrument positions, then the arithmetic mean is taken for all the individual declinations. In case the number of observations at the two positions is not equal, it is necessary to take into account the quantity $\frac{1}{2}(\delta_I - \delta_{II})$ for each observed declination, and only after this to calculate the mean declination.

In order to evaluate the accuracy of the resulting catalog, the mean square error of one observation is calculated. Sometimes the mean square error of the average declination is calculated for each star separately. A comparison of the obtained catalog with other catalogs enables one to judge its quality.

CHAPTER IX

The Absolute Determination of Right Ascensions

60. Principles of the Determination of Absolute Right Ascensions

In order to determine the right ascensions, the star transits are observed with a transit instrument or a meridian circle. The initial formula for the determination of absolute right ascensions is the Mayer formula:

$$\alpha_i = T_i + u_i + I \cdot i + K \cdot k + C \cdot c.$$

The transit times T_i are derived directly from observations according to the method described in Chapter VI. The coefficients of this formula

$$I = \cos z_m \sec \delta, \quad K = \sin z_m \sec \delta \text{ and } C = \sec \delta$$

can be calculated from the approximately known coordinates. In doing this, the principle of absolute determination is not violated since the inaccuracy of the quantities I, K, and C does not affect the determined right ascensions, since i, k and c are small (see Section 12).

The absolute observations are more complicated than the relative ones since it is necessary to determine and check the absolute parameters of the instrument mounting relative to the meridian and the vertical direction with special observations and readings of control instruments. The quantities i and c can be determined accurately by laboratory methods without any astronomical observations. Repeated determination of i and c over several intervals of time permits study of their variations and interpolation of the i and c values at the moment of stellar observation with sufficient accuracy.

The inclination of the horizontal axis and the collimation may be determined relatively simply. To determine the absolute azimuth of the instrument by the stars is somewhat difficult and requires an extended period of time. Therefore it is convenient to proceed from determining the rapidly varying instrument azimuth to a more steady azimuth of some direction fixed on the surface of the earth. For this purpose, the azimuth k of the horizontal axis of the instrument is divided into two components: 1) the azimuth of the horizontal axis relative to some fixed direction tied to earth landmarks (mires) and designated by k', and 2) the azimuth of this direction, the mire line Δk. The azimuth of the instrument k' relative to the mire line is easily determined by setting the movable vertical thread of the instrument on the landmarks, and can be taken into account. The quantity Δk can be determined only from astronomical observations. This convenient division of the instrument azimuth into its two components facilitates the control of its changes.

If the observed time T_i is corrected for the error in the orientation of the horizontal instrument axis and for the collimation (the diurnal aberration is assumed to be taken into account), then the required right ascension of the star will be connected with T_i through the relation $\alpha_i = T_i + u_i$, where u_i is the clock correction. The clock correction varies in the course of an evening of observation, hence the rate of the clock must be eliminated.

The diurnal rate ω of the clock is determined from observations of consecutive culminations of the same southern stars at twenty-four hour intervals. The transits of one and the same star at consecutive culminations differ from each other (after taking into account the instrumental errors and expressing the transits in terms of right ascensions) only by a change in the clock correction for the twenty-four hour period. It is even simpler to determine the clock rate by receiving radio time signals. Assuming that the run of the clock within twenty-four hour periods remains constant, it is possible to represent the clock correction for the time T_i in the form

$$u_i = u_0 + \frac{\omega}{24}(T_i - T_0),$$

where T_0 is some average time for a given evening, and u_0 is the corresponding clock correction. The times T_0 and T_i are here expressed in hours and fractions of hours.

After eliminating u_i, we obtain the relation

$$\alpha_i = T_i + u_0, \tag{30}$$

where T_i is corrected for the rate of the clock and u_0 is the clock correction for some average time T_0. Consequently, for every observing night, we have a system of n equations with $n + 1$ unknowns: n required right ascensions α_i and the clock correction u_0.

Under present conditions, each new series of observations is undertaken in order to increase the accuracy of the coordinates obtained from preceding determinations. Therefore it is possible to represent the true right ascension of the star in the form

$$\alpha_i = \alpha_i' + \Delta\alpha_i + \Delta A,$$

where α_i is the accepted right ascension taken from the improved catalog; $\Delta\alpha_i$ is the correction for α_i', which is unique for each star; ΔA is a common correction for all right ascensions in the catalog and is referred to as the equinox correction of the catalog. Then the relations (30) for any two stars is written as

$$\alpha_i' + \Delta\alpha_i + \Delta A = T_i + u_0,$$
$$\alpha_j' + \Delta\alpha_j + \Delta A = T_j + u_0;$$

subtracting, we obtain:

$$\Delta\alpha_i - \Delta\alpha_j = (T_i - T_j) - (\alpha_i' - \alpha_j').$$

Similar equations can be constructed from any combination of stars observed during a given night. In practice, these relations are usually formed for stars whose differences in right ascension do not exceed six hours.

Since the equations connecting the unknown quantities $\Delta\alpha_i$ tend to be scattered, the determination of these unknowns is possible only with an accuracy to within a constant. In order to obtain a definite solution, it is necessary to impose some condition. $\Sigma\Delta\alpha = 0$ is usually taken as such a condition for all the observed stars. Hence, the task of determining $\Delta\alpha_i$ is referred to as the smoothing of the right ascension system within itself. As a result of the solution of this problem one obtains improved right ascensions calculated from an erroneous point of vernal equinox of the catalog from which the α_i are taken. The displacement of this point relative to the true position of the vernal equinox is equal to ΔA. Since the position of the vernal equinox is defined by the intersection of two great circles, the true equator and the ecliptic, the determination of the quantity ΔA is in principle possible only from observations of the sun or other bodies of the solar system. The classical method for the determination of ΔA employs observations of the sun.

Thus, in order to determine the absolute right ascensions from observation, it is necessary to solve three problems: 1) to determine the azimuth of the horizontal axis of the instrument; 2) to smooth out the right ascension system within itself; 3) to determine the equinox correction. Let us consider the solution of these three basic problems.

61. Determination of the Relative Azimuth of the Instrument

The problem of determining the azimuth k of the instrument consists of two problems: determination of the instrument azimuth relative to some direction k', and the determination of the absolute azimuth Δk of this direction.

A fixed direction on the surface of the earth can be easily established by setting up along the meridian of the instrument a luminous sign or mark clearly visible within the field of view of the instrument. The establishment of such a meridian sign at a great distance from the instrument is almost always difficult because of local conditions. Therefore, the so-called mires proposed and first perfected by V. Struve at the Pulkovo Observatory found widespread application.

The mires (Fig. 55) are placed on brick (sometimes stone) pillars, along the instrument meridian, inside special booths at a distance of about 100 m from the meridian circle. This kind of mire has either a glass plate with an engraved line grid or a metallic plate with small holes to obtain starlike images. Behind the plate is placed a source of light. In order to be able to see the mire through the instrument, a lens having a focal length equal to its distance from the mire is placed on a special pillar inside the pavilion near the instrument. The mire lenses are sometimes placed on the same pillar with the meridian collimators or on special pillars outside the pavilion. In view of the long focal length of the mire lenses, it is not necessary that they be compound lenses. For this purpose, it is sufficient to use a single simple spherical lens.

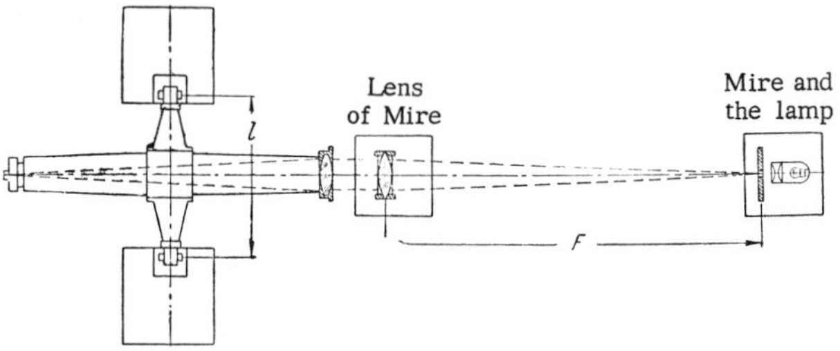

Fig. 55. Scheme of the mounting of a mire.

The line joining the mire and the second principal point of the objective defines some direction in space from which the relative azimuth of the instrument k' can be read off. Ordinarily, two mires are set up, one to the north and the other to the south of the instrument, thus increasing the accuracy in the determination of k' and eliminating the effect of collimation on its determination. If the location permits, the mires are set up along the horizon to avoid having to take into account the inclination of the horizontal axis.

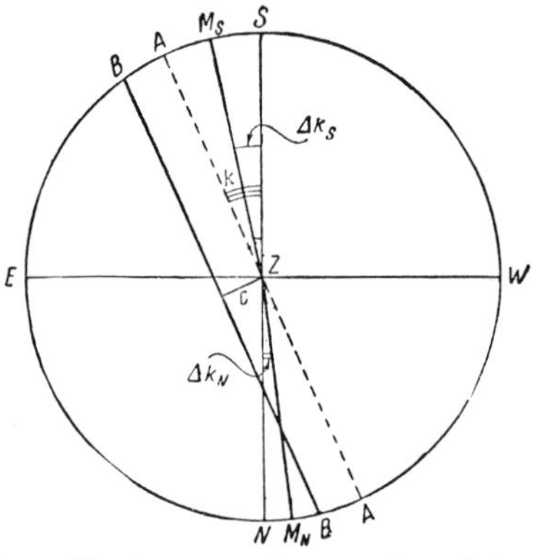

Fig. 56. Connection between the relative and absolute azimuths (angles in plane of horizon).

Figure 56 represents a projection of the celestial sphere (seen from above); *SN* is the true celestial meridian, *AA* is the great circle traced by the collimation-free line of sight of the instrument and inclined an angle k to the meridian. For a collimation c, the line of sight will trace the small circle BB parallel to AA. M_SZ and M_NZ are the directions defined by the southern and northern mires. The angles M_SZS and M_NZN, which we call the azimuths Δk_S and Δk_N of the mires, are respectively equal to

$$\begin{aligned} \Delta k_N &= \smile{A_N N} - \smile{A_N M_N} = k - k'_N, \\ \Delta k_S &= \smile{A_S S} - \smile{A_S M_S} = k - k'_S. \end{aligned} \tag{31}$$

If m_S and m_N are the screw readings in right ascension of the eyepiece micrometer of the instrument when set on the southern and northern mires, and m_c the collimation-free micrometer reading, then the relative azimuth of the instrument with respect to the northern and southern mires will be respectively equal to

$$\begin{aligned} k'_N &= (m_N - m_c) R_\alpha = m_N R_\alpha - c, \\ k'_S &= (m_c - m_S) R_\alpha = c - m_S R_\alpha, \end{aligned} \tag{32}$$

where R_α is the value of one screw revolution in right ascension.

From the equations (31) and (32), the instrument azimuth will be equal to

$$k = \frac{(m_N - m_S) R_\alpha}{2} + \frac{\Delta k_S + \Delta k_N}{2} = k' + \Delta k.$$

The bisectrix of the angle between the two lines of sight of the mires projected to the horizon is called the middle line of the mires. Then the first term k' is the azimuth of the instrument relative to this middle line, or the *relative azimuth*. If the mires are accurately set up, then k' will be nearly equal to the true azimuth k, and we can use it as the first approximation of the instrument azimuth in the calculations. The second term is the *absolute azimuth* Δk of the middle line of the mires, which is determined by special astronomical observations.

In determining the relative azimuth of the instrument systematically, it is possible to assume that its changes during observations are equal to the changes of the absolute azimuth of the instrument. This is possible since the azimuth of the middle line of the mires is much more stable than the azimuth of the horizontal axis of the instrument and can be considered practically constant during a certain period of time. Actually, the stability of the azimuths of the instrument and the mires is determined by the stability of the instrument supports and of the lens of the mire and the mire itself. The maximum change in the quantities k' and Δk occurs if, correspondingly, the pillars of the instrument shift in the direction of the meridian, or the pillar of the mire in the direction perpendicular to it. However, equal shifts Δ of the pillars will cause different changes in the azimuth: $\frac{\Delta}{l}$ is the change in the instrument azimuth (l is the distance between the working sections of the pivots of the instrument) and $\frac{\Delta}{F}$ is the change in the azimuth of the mire (F is the focal length of the mire lens). Therefore, the azimuth of the mires is more stable than the azimuth of the instrument by the ratio $\frac{F}{l}$; that is, approximately 100 times. The changes in the position of the middle line of the mires are usually so slow that it is possible to assume that its azimuth is constant during a period of 10-15 days.

62. Determination of the Azimuth of the Middle Line of Mires

The azimuth of the middle line of the mires is determined through observations of circumpolar stars at two culminations. According to the Pulkovo method, only one polar star (Polaris or α Ursae Minoris) is observed for this purpose. This star is sufficiently bright so that it may be observed not only at night but also during the day. Observations of both culminations of Polaris are made during the whole course of catalog observations. During each observation of Polaris the micrometer is read when pointing at the mire and the transit of any chosen clock star is observed in order to determine the rate of the clock.

By accepting in the first approximation the relative azimuth k' as the true azimuth of the instrument, one can calculate the approximate right ascension of Polaris:

$$\alpha' = T + u' + Kk',$$

where T is the transit time modified by the known corrections, and u' is the accepted approximate clock correction.

The true right ascension of Polaris can be represented in the form

$$\alpha = \alpha' + K\Delta k + \Delta u,$$

where Δu is the correction of the accepted value u'. On the other hand, $\alpha = \alpha_e + \Delta\alpha$, where α_e is the ephemeris right ascension taken from the almanac and $\Delta\alpha$ is its correction. From a combination of these two expressions, we obtain

$$K\Delta k - \Delta\alpha + \Delta u = \alpha_e - \alpha',$$

where $\alpha_e - \alpha' = D$, which is a known quantity.

Two such equations:

$$\begin{aligned} K_V\Delta k - \Delta\alpha + \Delta u &= D_V \text{ for the upper culmination} \\ K_N\Delta k - \Delta\alpha + \Delta u &= D_N \text{ for the lower culmination} \end{aligned} \tag{33}$$

determine the unknown Δk under the assumption that it is constant. Since the quantities K_V and K_N are large for Polaris, mistakes in the determination of the rate of the clock do not play an essential part. The rate of the clock is assumed eliminated by observing a southern star or upon receiving radio time signals, and the quantity Δu may be assumed to be the same in both equations. The quantity $\Delta\alpha$ is also constant since the effect of precession, nutation, annual aberration and the proper motion are taken into account in the almanac.

Subtracting one equation (33) from the other, we obtain

$$\Delta k = \frac{D_V - D_N}{K_V - K_N},$$

where

$$K_V = \sin(\varphi - \delta)\sec\delta, \quad K_N = \sin(\varphi + \delta)\sec\delta.$$

Coefficients K_V and K_N in the Mayer formula have different signs, since $\delta > \varphi$ for Polaris. Consequently, the difference in the denominator is in fact a sum and the accuracy of the determination of the azimuth is significantly higher than the accuracy in recording the transit of Polaris.

The mean square error of determining the azimuth $\varepsilon_{\Delta k}$ can be represented in the following form:

$$\varepsilon_{\Delta k} = \pm \frac{\varepsilon_{D_V - D_N}}{K_V - K_N} = \pm \frac{\varepsilon_\alpha \sqrt{2}}{K_V - K_N},$$

where ε_α is the mean square error of the right ascension and depends basically on the accuracy of the time recording. For Polaris, this quantity, derived from the agreement of the determinations of its right ascension, is close to $\pm 0^s.5$-$\pm 0^s.6$. Since for the latitude of Moscow K_V = -34 and K_N = 36, the accuracy of one determination of the azimuth Δk from observations of Polaris at two culminations is equal on the average to $\varepsilon_{\Delta k} = \pm 0^s.011$.

In principle, it is possible to determine the azimuth of the middle line of the mires Δk by combining each pair of observations in the upper and lower culminations. However, it is much more economical and simpler to average out the observations for a certain period of time for the upper and lower culminations separately. We thus obtain the average azimuth Δk for this period under the assumption that Δk has not changed. When choosing the period for averaging the observations, it is necessary to consider the stability of the middle line of the mires: any change in Δk within this period must be negligibly small. In practice, Δk is constant to within $0^s.01$ for a half-month period if the mire lenses have a focal length of about 100 meters.

Changes in the azimuth of the middle line of the mires occur smoothly and slowly, as has been shown by observations, and essentially follow the annual temperature changes. Variations with periods of one year or twenty-four hours appear because the ground freezes and because of the direct influence of the temperature changes, causing deformation and displacement of the mire lens supports and of the mire itself. Placing heat-insulating sheaths around the supports and setting them to a depth below the freezing level increase the smoothness of yearly and diurnal changes and decrease the diurnal amplitude of the quantity Δk. The long-term shift of the middle line of mires due to geological displacements of the soil and structural changes in the foundation layers is small.

The observed quantities Δk also change because of the shifting of the poles of the earth on its surface. This causes a change in the position of the meridian which determines the zero-point reading of the azimuth.

Figure 57 shows the curve representing the variation of the azimuth of the middle line of the mires for the transit instrument of the Pulkovo Observatory as obtained from observations of Polaris in the period between 1920 and 1925.

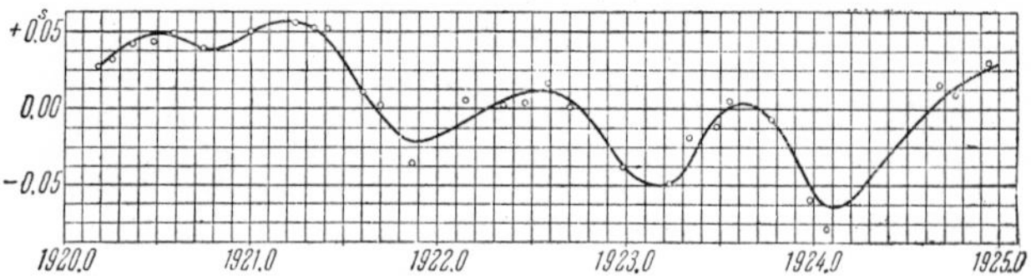

Fig. 57. Variation of the azimuth of the middle line of mires.

Several methods have been proposed for checking the stability of the mire line. Thus, for example, at the Cape Observatory a mercury horizon was placed at the rocky bottom of a well dug near the mire. This provided a stable vertical direction and made it possible to measure the shifts of the mires relative to it.

63. Determination of the Absolute Supplementary Orientation of the Horizontal Axis

Frequently, along with Δk, the quantity Δi, called the side flexure, is found. The term "side flexure" is somewhat conditional since the side refraction, the change in collimation with elevation, etc., also contribute to it. It is possible to determine Δi since the observations, as a rule, are made at two instrument positions. It is assumed that the sign of Δi changes when the instrument is reversed. The correction for the side flexure at the time of observation can be represented by the relation

$$I\,\Delta i = \Delta i \cos z_m \sec \delta,$$

assuming that its magnitude varies with the zenith distance according to the cosine law. Therefore, Δi is sometimes called the supplementary inclination; similarly, the quantity Δk is called the supplementary azimuth, whereas the determination of both these quantities is referred to as the determination of the supplementary orientation of the horizontal axis.

Thus the equations for each circle position during the observation of the upper and lower culminations may be written in the following form:

$$\left.\begin{aligned} K_V \Delta k + I_V \Delta i - \Delta\alpha - \Delta u &= D_V^W \\ K_N \Delta k + I_N \Delta i - \Delta\alpha - \Delta u &= D_N^W \end{aligned}\right\} \text{for the instrument position circle-west}$$

$$\left.\begin{aligned} K_V \Delta k - I_V \Delta i - \Delta\alpha - \Delta u &= D_V^E \\ K_N \Delta k - I_N \Delta i - \Delta\alpha - \Delta u &= D_N^E \end{aligned}\right\} \text{for the instrument position circle-east}$$

A system of such equations is made up for the period of time during which it is possible to assume the quantity Δk to remain constant. The solution of this system will determine Δk and Δi in an absolute way.

The Pulkovo astronomer, A. A. Nemiro, proposed a better method for processing the observations in order to derive the supplementary orientation of the horizontal axis. This method was successfully applied in the revision of all Pulkovo Observatory catalogs of right ascensions.

Let us write down the basic equation for observation of a circumpolar star in the form:

$$\alpha_e + \Delta\alpha = T + u + I\,\Delta i + K\,\Delta k.$$

Besides observations of a circumpolar star (sometimes several), the analysis includes observations of a number of clock stars, for which the basic equations are averaged out:

$$\overline{\alpha}_{ei} + \overline{\Delta\alpha}_i = \overline{T}_i + u + \overline{I}_i\,\Delta i + \overline{K}_i\,\Delta k.$$

Subtracting from the equation for the circumpolar star the average equation for the group of clock stars and designating the known quantity $\alpha_e - T - [\overline{\alpha}_{ei} - \overline{T}_i]$ by l, we have:

$$(K - \overline{K}_i)\,\Delta k + (I - \overline{I}_i)\,\Delta i - (\Delta\alpha - \overline{\Delta\alpha}_i) = l.$$

Let us use this equation for the observations of a circumpolar star in two culminations and in two instrument positions. Neglecting the quantity $\overline{\Delta\alpha}_i$ since it is small in comparison with $\Delta\alpha$, we obtain a system of equations

$$\left.\begin{aligned} (K_V - \overline{K}_i)\,\Delta k + (I_V - \overline{I}_i)\,\Delta i - \Delta\alpha &= l_V^W, \\ (K_N - \overline{K}_i)\,\Delta k + (I_N - \overline{I}_i)\,\Delta i - \Delta\alpha &= l_N^W, \\ (K_V - \overline{K}_i)\,\Delta k - (I_V - \overline{I}_i)\,\Delta i - \Delta\alpha &= l_V^E, \\ (K_N - \overline{K}_i)\,\Delta k - (I_N - \overline{I}_i)\,\Delta i - \Delta\alpha &= l_N^E. \end{aligned}\right\} \quad (34)$$

Since the number of observations for each position of the instrument may be different, the solution of this system by the method of least squares is made by assigning weights proportional to the number of observations of a circumpolar star. The quantity $\Delta\alpha_e$ is obtained with small weight but this is unimportant.

Equations (34) can be significantly simplified by using the Bessel formula; for this purpose, we represent the coefficients of the Mayer formula by the expressions

$$I_V = \cos\varphi + \sin\varphi\tan\delta, \quad K_V = \sin\varphi - \cos\varphi\tan\delta,$$
$$I_N = \cos\varphi - \sin\varphi\tan\delta, \quad K_N = \sin\varphi + \cos\varphi\tan\delta.$$

After this, equations (34) take the form

$$(\tan\delta_V - \overline{\tan\delta}_i)(\sin\varphi\,\Delta i - \cos\varphi\,\Delta k) - \Delta\alpha = l_V^W,$$
$$(\tan\delta_N - \overline{\tan\delta}_i)(-\sin\varphi\,\Delta i + \cos\varphi\,\Delta k) - \Delta\alpha = l_N^W,$$
$$(\tan\delta_V - \overline{\tan\delta}_i)(-\sin\varphi\,\Delta i - \cos\varphi\,\Delta k) - \Delta\alpha = l_V^E,$$
$$(\tan\delta_N - \overline{\tan\delta}_i)(\sin\varphi\,\Delta i + \cos\varphi\,\Delta k) - \Delta\alpha = l_N^E.$$

Since, according to the adopted condition, Δi changes its sign when we go from one instrument position to the other, then

$$\Delta n_W = \Delta i\sin\varphi - \Delta k\cos\varphi \text{ and } \Delta n_E = -\Delta i\sin\varphi - \Delta k\cos\varphi.$$

Therefore

$$(\tan\delta_V - \overline{\tan\delta}_i)\,\Delta n_W - \Delta\alpha = l_V^W,$$
$$(\tan\delta_N - \overline{\tan\delta}_i)\,\Delta n_W - \Delta\alpha = l_N^W,$$
$$(\tan\delta_V - \overline{\tan\delta}_i)\,\Delta n_E - \Delta\alpha = l_V^E,$$
$$(\tan\delta_N - \overline{\tan\delta}_i)\,\Delta n_E - \Delta\alpha = l_N^E.$$

From this we find for each instrument position:

$$\Delta n_W = \frac{l_V^W - l_N^W}{2\tan\delta} \text{ and } \Delta n_E = \frac{l_V^E - l_N^E}{2\tan\delta},$$

since $\tan\delta_V = -\tan\delta_N$. Corrections for the observations of the supplementary azimuth and the supplementary inclination can be found from the formulas:

$$\Delta T_W = \pm\, n_W \tan\delta \text{ and } \Delta T_E = \pm\, \Delta n_E \tan\delta,$$

where the positive sign corresponds to the upper culmination and the negative to the lower. If it is necessary to know Δk and Δi, they can be found by using the formulas

$$\Delta k \cos\varphi = -\frac{1}{2}(\Delta n_E + \Delta n_W) \text{ and } \Delta i \sin\varphi = \frac{1}{2}(\Delta n_W - \Delta n_E).$$

The advantage of this method for determining the corrections of the supplementary orientation of the horizontal axis comes from the possibility of determining them from observations at each instrument position separately, which enables one to use shorter series of observations. Moreover, notice the simplicity of the calculations.

64. Equalizing the Right Ascension System

After taking into account the instrument azimuth determined by the absolute method as well as the side flexure (if it has been determined) the equation for the determination of the right ascension of each star observed during a certain evening will have the form:

$$\alpha_i = T_i + u_i.$$

Taking into account the run of the clock, it is possible to reduce the unknown quantities, that is, the clock corrections u_i for different moments of time, to one unknown u_0 pertaining to some average moment of time. Assuming that the correction for the run of the clock is implicit in T_i (i.e., $\alpha_i = T_i + u_0$) and designating by u_i' the clock correction calculated for the transit time T_i and with the accepted value of the right ascension α_i' taken from the catalog (i.e., $u_i' - \alpha_i' = T_i$), we have:

$$\alpha_i - \alpha_i' = u_0 - u_i' \text{ or } \Delta\alpha_i + \Delta A = u_0 - u_i'.$$

The unknowns u_0 and ΔA cannot be determined independently since they enter into each equation with equal coefficients. Therefore we drop in what follows the unknown ΔA from the equation and we shall assume that it is included in the quantity u_0.

However, the determination of the quantities $\Delta\alpha_i$ from the equations is not possible without imposing on them some condition, since the n equations corresponding to observations of n stars connect $n + 1$ unknowns: n unknowns $\Delta\alpha_i$ and u_0. Which condition is imposed has no essential importance. All the $\Delta\alpha_i$ calculated for one condition will differ by a constant quantity ΔN from all $\Delta\alpha_i$ calculated with another condition. In other words, the zero points of two catalogs, formed by adding to the quantities α_i' and α_i'' the corrections $\Delta\alpha_i'$ and $\alpha\Delta_i''$ corresponding to the two systems, will also differ by the quantity ΔN. It will enter into the equinox correction ΔA of the catalog and will be determined together with it. If we impose the condition $\sum\Delta\alpha_i = 0$ on all stars of the catalog, then the zero point of the new, improved catalog will coincide with the zero point of the original one; that is, their equinox corrections will be equal. For each night of observing, we have n equations of the form

$$\Delta\alpha_i = u_0 - u_i'. \tag{35}$$

The solution of these equations, that is, the equalization of the system of right ascensions, can be made by several methods. The principle of absolute determination can most easily be satisfied by the Pulkovo method of equalization. According to this method, the system of right ascensions is equalized by using only thirty-six bright stars distributed along the celestial equator ($|\delta| < 40°$) and easily accessible for observation at any time of night or day. Continuous observations of these stars, often referred to as the fundamental stars or Maskelyne stars, (after the English astronomer of the 18th century who composed the catalogs of these stars) during twenty-four hours establish the connection between solar observations and observations of the catalog stars. After smoothing out the system of right ascensions of the fundamental stars, the right ascensions of all other stars are obtained by a purely differential method. This method was first applied for reducing the Pulkovo catalogs in the years 1845 and 1865.

Taking the differences between the equations (35) formed for any two fundamental stars observed during one evening, so long as the interval between these observations does not exceed six hours, we eliminate the unknown u_0 and obtain for the total period of observations a fairly large group of equations of the form

$$\Delta\alpha_i - \Delta\alpha_j = u_j' - u_i'. \tag{36}$$

These equations are solved by the method of least squares under the assumption that $\sum\Delta\alpha_k = 0$. By this method the new system is attached to the point of the spring equinox of the reference catalog.

The solution of the system of equations (36) with thirty-six unknowns and with the number of equations reaching several thousand, is cumbersome and labor-consuming. Therefore the equations are sometimes averaged out within three-hour groups, and eight mean corrections are determined for these groups. Corrections for each right ascension are found from the subsystem of equations made up for the three-hour groups. Naturally, averaging out in three-hour groups leads to equalizing the magnitudes of the determined corrections $\Delta\alpha_i$. With the help of the obtained corrections $\Delta\alpha_i$, it is possible to determine for a certain evening (from observations of the fundamental stars) the quantity u_0 from the expression $u_0 = u_i' + \Delta\alpha_i$ and then, by applying $\Delta\alpha_j = u_0 - u_j'$, to find the corrections $\Delta\alpha_j$ for the remaining stars. Thus the right ascensions of stars are obtained from the equalized system of the fundamental stars. We must still determine the zero-point correction of the system, or the equinox correction.

When certain of the Pulkovo catalogs were developed, some additional terms of the form $a \sin t + b \cos t$ were introduced, where t is the mean time of the observations. Then equations (36) take on the form

$$\Delta\alpha_i - \Delta\alpha_j = u'_j - u'_i - a(\sin t_i - \sin t_j) - b(\cos t_i - \cos t_j).$$

The added members must take into account the systematic errors in the observations for the diurnal period: the so-called diurnal term. In view of the small size of the diurnal term and the complexity of its variation with t, the determination of the coefficients a and b is uncertain. In what follows, this simplifying method is ignored.

In comparison with the Pulkovo method, other methods for equalization have some fundamental disadvantages. For example, in the widely used Greenwich method the condition $\sum \Delta\alpha_i = 0$ for the solution of equations (35) is imposed for one observing night. This makes it possible, by summing up the equations, to determine the quantity $u_0 = \frac{1}{n}\sum u'_i$. However, this condition leads to the violation of the principle of absolute determination since the systematic errors of the original catalog are carried over, although in somewhat smoothed form, to the new right ascensions.

As a matter of fact, the original catalog may contain a systematic error varying in accordance with changing right ascensions. This error, designated by $\Delta\alpha_\alpha$, ordinarily varies roughly as a sinusoid (Fig. 58), as investigation has shown. This is so since the main causes appear to be the seasonal changes of the surroundings of the instrument. Since observations for one night include only several hours of right ascension, (as is shown, for example, in Fig. 58), then in forming $u_0 = \frac{1}{n}\sum u'_i$ some mean error $\overline{\Delta\alpha_\alpha}$ will be included in the hours of right ascension covered by the observations. Into the difference $u_0 - u'_i$ enters the quantity $\overline{\Delta\alpha_\alpha} - \Delta\alpha_\alpha$ instead of the quantities $\Delta\alpha_\alpha$ needed to compensate fully for the systematic errors in α'_i while working out the new right ascension: $\alpha_i = \alpha'_i + \Delta\alpha_i$. In this way, a more or less equalized systematic error α_α of the original catalog will enter into the new catalog, depending upon the duration of the nightly observations.

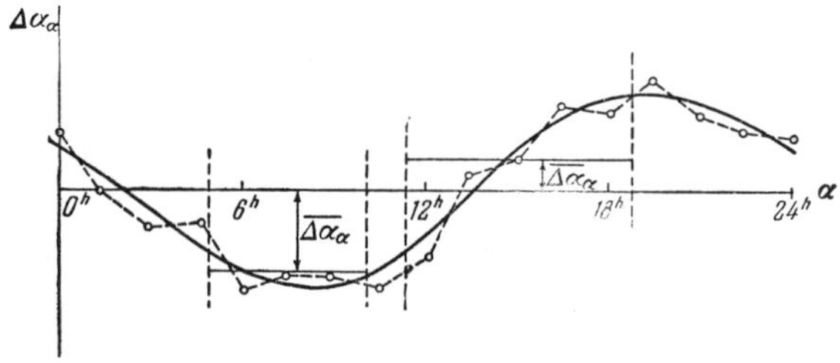

Fig. 58. Influence of the systematic errors of the type $\Delta\alpha_\alpha$ in the Greenwich method of equalization.

Among the newer methods of equalization there is the chain method proposed at Pulkovo. This method resembles the ordinary chain method of latitude observations with a zenith telescope.

The number of stars used for equalization is usually about 200; they are, as far as possible, distributed evenly along the equatorial zone (±25°), and are divided into 12 groups in two-hour intervals of right ascension. Each group is observed for two months: one month with the preceding group and one month with the following group. Thus, if in January groups I and II were observed, then in February groups II and III are observed, and in March III and IV, etc. As in every chain method of observation, it is recommended that groups be observed which culminate near midnight, when the observation conditions change only slightly. The rigid condition is the requirement to observe all the stars of both groups without exception.

Let us introduce the following small quantities: $\Delta u_0 = u_0 - \frac{1}{n}\sum u_i'$ and $\Delta u_i' = \frac{1}{n}\sum u_i' - u_i'$; the latter quantity is the deviation of the clock correction α_i' (calculated from u') from its mean value for the night. Then the basic equation $\Delta\alpha_i = u_0 - u_i'$ can be transformed into the form $\Delta\alpha_i = \Delta u_i' - \Delta u_0$. The unknown quantity Δu_0 is small since the average calculated clock correction $\bar{u} = \frac{1}{n}\sum u_i'$ differs only slightly from the quantity u_0. This is true since the errors of the original catalog (as a rule, not large) compensate for each other in the average during the night. Consequently, the known quantity $\Delta u_i'$ basically determines the magnitude of the correction $\Delta\alpha_i$, and thus makes it possible to exclude from the program stars requiring large corrections, which naturally improves the equalization procedure. Besides this, in comparing the quantities $\Delta u_i'$ for various nights, it is possible to exclude individual observations and nights which deviate significantly from average due to anomalous personal-instrumental errors and those evenings showing a progressive change of these errors with time.

In forming the average value $\Delta u_i'$ for each star for one month we obtain the equation

$$\Delta\alpha_i = \overline{\Delta u_i'} + \overline{\Delta u_0}, \tag{37}$$

where $\overline{\Delta u_0} = \frac{1}{n}\sum \Delta u_0$ is the average Δu_0 for n nights in the course of which stars of a given group were observed. We then proceed to the fictitious mean star of the group by averaging equations (37) for all stars of the given group separately for the first and second month of observations. Then for each month we obtain two such equations:

$$\overline{\Delta\alpha_j} = \overline{\overline{\Delta u_j'}} + \overline{\Delta u_0} \text{ and } \overline{\Delta\alpha_{j+1}} = \overline{\overline{\Delta u_{j+1}'}} + \overline{\Delta u_0},$$

where j is the group number. Since groups j and $j+1$ were observed in each month during the same nights, then for each such pair $\overline{\Delta u_0}$ is the same, and consequently in forming the differences it drops out. As a result we shall have the following system of equations:

$$\overline{\Delta\alpha}_2 - \overline{\Delta\alpha}_1 = \overline{\overline{\Delta u}}{}'_2 - \overline{\overline{\Delta u}}{}''_1,$$

$$\overline{\Delta\alpha}_3 - \overline{\Delta\alpha}_2 = \overline{\overline{\Delta u}}{}'_3 - \overline{\overline{\Delta u}}{}''_2,$$

$$\cdot\ \cdot\ \cdot\ \cdot\ \cdot\ \cdot\ \cdot\ \cdot\ \cdot\ \cdot\ \cdot\ \cdot$$

$$\overline{\Delta\alpha}_1 - \overline{\Delta\alpha}_{12} = \overline{\overline{\Delta u}}{}'_1 - \overline{\overline{\Delta u}}{}''_{12}.$$

The superscript $''$ indicates that the quantities $\overline{\overline{\Delta u}}{}''_i$ were obtained in a different month from $\overline{\overline{\Delta u}}{}'_i$. These equations are easily solved when the condition $\sum \overline{\Delta\alpha}_j = 0$ is imposed.

In determining $\Delta\alpha_j$, we find the quantities $\overline{\Delta u_0} = \overline{\Delta\alpha_j} - \overline{\overline{\Delta u_j}}{}'$ and the quantities $\Delta\alpha_i$ from $\Delta\alpha_i = \overline{\Delta u'_i} + \overline{\Delta u_0}$. Substituting them in $\Delta\alpha_i = \Delta u'_i + \Delta u_0$, it is possible to derive the quantity Δu_0 for the night and with it determine $\Delta\alpha$ for the remaining stars.

A disadvantage of the method is the necessity to observe all the stars of both groups during the same evening, which is not possible for all observatories due to meteorological conditions. If there are individual evenings with only a few stars left out, they can all be used by reducing them, according to the $\Delta u'_i$ differences for the stars generally, to the system of a complete evening. The advantage of the method is the simplicity of the analysis, with a large number of observed stars being used to equalize the right ascension system.

65. Determination of the Equinox Correction from Observations of the Sun

In order to determine the equinox correction, it is necessary to observe objects of the solar system simultaneously with the catalog stars. Observations of the sun are used in the classical method. The sun's declinations are observed by the absolute method, and the right ascension is obtained in the system of the improved catalog. If we know the declination of the sun, we may calculate its true right ascension. By comparing the latter with the observed right ascension of the sun in the system of the catalog to be improved, it is possible to derive the equinox correction ΔA. In order to connect the observations of the sun with the observations of stars, Maskelyne stars are used.

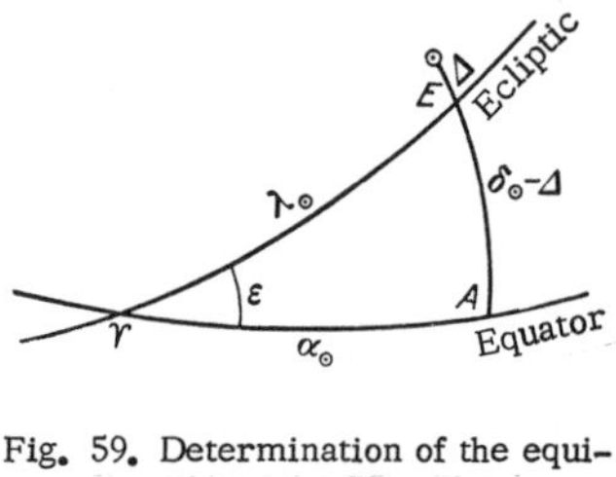

Fig. 59. Determination of the equinox correction.

Let Fig. 59 represent the position of the sun relative to the point of spring equinox. Generally speaking, the sun is not on the ecliptic but at some latitude $b_{\odot}$. From the triangle ΥAE, it follows that $\tan(\delta_{\odot} - \Delta) = \sin\alpha_{\odot} \tan\varepsilon$, where $\Delta = \odot E$.

If the quantities α_e, δ_e, ε_e designate, for example, values taken from an astronomical almanac, then by considering the quantities $\Delta\delta_\odot = \delta_\odot - \delta_e$, $\Delta\alpha_\odot = \alpha_\odot - \alpha_e$, $\Delta\varepsilon = \varepsilon - \varepsilon_e$ to be the final increments, it is possible to write down the following formula:

$$\Delta\delta_\odot \sec^2 \delta_e = \Delta\alpha_\odot \cos \alpha_e \tan \varepsilon_e + \Delta\varepsilon \sin \alpha_e \sec^2 \varepsilon_e. \quad (38)$$

The quantity $\Delta\alpha_\odot$ which is the difference between the true and the ephemeris right ascension of the sun can be transformed in the following way:

$$\Delta\alpha_\odot = \alpha_\odot - \alpha_e = \alpha + \Delta A - \alpha_e = \Delta\alpha + \Delta A,$$

where α is the right ascension of the sun in the system of the catalog to be improved, which has an error ΔA, and $\Delta\alpha = \alpha - \alpha_e$.

Consequently, by grouping the unknowns on the left hand side, relation (3) may be written in the following form:

$$\Delta A \cos \alpha_e \tan \varepsilon_e + \Delta\varepsilon \sin \alpha_e \sec^2 \varepsilon_e = \\ = \Delta\delta_\odot \sec^2 \delta_e - \Delta\alpha \cos \alpha_e \tan \varepsilon_e. \quad (39)$$

The quantities $\Delta\delta_\odot$ and $\Delta\alpha$ are derived from a comparison of the observed absolute declinations and the relative right ascensions of the sun with their ephemeris values. Equations (39) can be written for each observation of the sun for any number of years. Solving them by the method of least squares, it is possible to find the required equinox correction ΔA and also $\Delta\varepsilon$.

Bessel introduced into the left hand side more unknowns, accounting for the effect of systematic errors on the observed declination of the sun: the inaccuracy of the accepted latitude $\Delta\varphi$, the corrections for the refraction coefficient $\Delta\rho$ and the instrumental errors depending on the zenith distance Δz. He then represented the quantity $\Delta\delta_\odot$ in the form:

$$\Delta\delta_\odot = (\delta_\odot - \delta_e) + \Delta\varphi + \Delta\rho \tan z + \Delta z \sin z.$$

Thus, the equation (39) has been transformed into:

$$a\,\Delta A + b\,\Delta\varepsilon + c\,\Delta\varphi + d\,\Delta\rho + e\,\Delta z = \Delta\delta_\odot \sec^2 \delta_e - \Delta\alpha \cos \alpha_e \tan \varepsilon_e$$

where the coefficients of the terms in the left side are known.

This method was used at Pulkovo until the year 1900. The disadvantage of the method is that in solving the equations, both the declination and the right ascension of the sun are used together. The systematic errors of declinations necessitate the introduction of a large number of unknowns, which decreases the weight of the unknown ΔA.

A better method is the Newcomb method, which is at present replacing the previous method of analysis. In the Newcomb method, $\delta_\odot$ and $\Delta\alpha$ are used separately, and all calculations are made for determination of ΔA.

The true right ascension of the sun may be represented in the form:

$$\alpha_\odot = \alpha + \Delta A = \alpha_e + \Delta\alpha_e,$$

where α is the observed right ascension in the system of the improved catalog, and $\Delta\alpha_e$ is the correction for the ephemeris value. In designating the quantity $\alpha - \alpha_e$ determined through the observations by $\Delta\alpha$, we have:

$$\Delta A = \Delta\alpha_e - \Delta\alpha. \tag{40}$$

The quantity $\Delta\alpha_e$ is determined from the absolute observations of the declinations of the sun. The quantity ΔA can be found from observations averaged over the full year provided $\overline{\Delta\alpha}_e$, the average correction for the year, is known. The quantity is close to the quantity $\overline{\Delta\lambda}_e$, where λ_e is the longitude of the sun taken from the almanac. It can be proved that $\overline{\Delta\alpha}_e : \overline{\Delta\lambda}_e = 1$.

Indeed, we have $\tan\alpha = \cos\varepsilon \tan\lambda$ from the spherical triangle ΥAE (Fig. 59). By differentiating and then replacing the differentials by finite differences, we obtain:

$$\frac{\Delta\alpha}{\cos^2\alpha} = \Delta\lambda \cos\varepsilon \frac{1}{\cos^2\lambda} - \Delta\varepsilon \sin\varepsilon \tan\lambda$$

or

$$\Delta\alpha = \cos\varepsilon \frac{\cos^2\alpha}{\cos^2\lambda} \Delta\lambda - \sin\varepsilon \tan\lambda \cos^2\alpha\, \Delta\varepsilon. \tag{41}$$

Let us show that on the average, during the year, the coefficients of $\Delta\lambda$ and $\Delta\varepsilon$ are nearly equal, respectively, to one and zero. For this we shall use the relations obtainable from the same triangle:

$$\tan\alpha = \cos\varepsilon \tan\lambda,$$
$$\cos\alpha \cos\delta = \cos\lambda,$$
$$\sin\varepsilon \sin\lambda = \sin\delta.$$

We transform (41) by substituting $\frac{\tan\alpha}{\tan\lambda}$ for $\cos\varepsilon$ in the first coefficient and $\frac{\sin\delta}{\sin\lambda}$ for $\sin\varepsilon$ and $\frac{\cos\lambda}{\cos\delta}$ for $\cos\alpha$ in the second coefficient. We then have

$$\Delta\alpha = \frac{\tan\alpha}{\tan\lambda} \frac{\cos^2\alpha}{\cos^2\lambda} \Delta\lambda - \frac{\sin\delta}{\sin\lambda} \frac{\sin\lambda}{\cos\lambda} \frac{\cos\lambda}{\cos\delta} \cos\alpha\, \Delta\varepsilon$$

or

$$\Delta\alpha = \frac{\sin 2\alpha}{\sin 2\lambda} \Delta\lambda - \tan\delta \cos\alpha\, \Delta\varepsilon.$$

On integrating over the whole year, the quantity $\frac{\sin 2\alpha}{\sin 2\lambda}$ is equal to one, and $\tan\delta \cos\alpha$ is equal to zero. Thus, under the condition that one uses equation (41) for one or more years, it is possible to replace $\Delta\alpha_e$ by $\Delta\lambda_e$.

The quantity $\overline{\Delta\lambda}_e$ is determined from observations of declinations of the sun in the following way. Let us take the derivative of the expression $\ln\sin\delta = \ln\sin\varepsilon + \ln\sin\lambda$:

$$\frac{\Delta\delta}{\tan\delta} = \frac{\Delta\varepsilon}{\tan\varepsilon} + \frac{\Delta\lambda}{\tan\lambda}.$$

Substituting the relations $\tan\delta = \sin\alpha \tan\varepsilon$ and $\tan\lambda = \tan\alpha/\cos\varepsilon$ obtained from the triangle ΥEA, we have, after a transformation: $\Delta\delta = \Delta\varepsilon \sin\alpha + \Delta\lambda \sin\varepsilon \cos\alpha$. Introducing the constant quantity $\Delta\delta_{\odot} = x$, and designating $\Delta\varepsilon$ by y and $\Delta\lambda \sin\varepsilon$ by z, we obtain the fundamental equation in the form:

$$\Delta\delta = x + y \sin\alpha + z \cos\alpha.$$

The quantity $\Delta\delta = \Delta\delta_{\odot} = \delta_{\odot} - \delta_e$ is obtained from observations.

Averaging such equations for each month, we obtain a system of twelve conditional equations for each year. Solving by the method of least squares, we obtain z; we then find

$$\overline{\Delta\lambda}_e = \overline{\Delta\alpha}_e = \frac{z}{\sin\varepsilon}.$$

Knowing the quantity $\overline{\Delta\alpha}_e$, we find the equinox correction ΔA in which we are interested from relation (40).

Observations of the sun and the fundamental stars made in day time have systematic errors which differ from the errors in the observations of catalog stars made during the night. Besides this, the accuracy of the derived coordinates of the sun is considerably lower than the accuracy of star observations since the settings are made at the edge of the sun's disk, which introduces large personal errors.

In place of the sun, one may observe any planet of the solar system. The position of the ecliptic and consequently of the point of the vernal equinox is obtained in this case indirectly by determining the relative positions of the orbits of the planet and of the earth. This does not present any difficulties for the well-constructed theories of the motion of the observed planets. Observation of such bodies as the moon and the large planets (Mercury, Venus, Mars), in fact, removes the problem of the systematic difference between night and day observations; however, the necessity of removing personal errors arising from settings made on the edges of the disks of the luminous bodies remains, as well as the necessity to take into account the phase error. These difficulties disappear in observing minor planets (see Section 81). At present the equinox correction ΔA is not determined for individual catalogs. Series of observations of the sun and the planets extending over several years and carried out at different observatories from time to time, are taken together and used for computing the equinox correction only for the fundamental catalog.

CHAPTER X

Photographic Method for Determining Relative Coordinates and Proper Motions

66. Application of the Photographic Method in Astrometry

The development of astronomy in studying the structure of the universe and, in the first place, of the galaxy has presented the astrometrists with the problem of determining the positions and proper motions for a large number of stars within the system of some fundamental catalog. This task can be and has, to some extent, been solved in the 19th century by visual means; however, it can be more conveniently and precisely done by photographic methods.

The photographic method is considered important in astrometry, in the first place, because it fixes on a photographic plate the relative positions of the stars and other celestial objects in some region of the sky and conserves this picture for future investigations. The second advantage of the photographic method is the large number of objects observed at one time. A great many star images are obtained on a negative, sometimes up to several thousands, as well as of other celestial objects, among which the extragalactic nebulae are of first importance. And, finally, the third advantage is that fainter objects are reached by the photographic method than by visual observations. It is necessary to point out that excessive brightness causes difficulties for photographic observations. The dividing line separating the usefulness of the visual and photographic methods is at approximately the ninth stellar magnitude.

There is a limitation in the photographic method caused by the fact that the plate contains an image of an isolated portion of the sky. Therefore one can study only the positions of stars relative to each other, or, by comparing several plates obtained at different epochs, one can study only the relative displacements of the stars. The position of the coordinate network on the plate can be determined only from known coordinates of reference stars.

Photographic astrometry is a division of astrometry which applies photography to the solution of astrometric problems; these include the problems of determining the relative coordinates and proper motions for a large number of objects as well as for individual celestial objects. It is also concerned with such problems as the determination of parallaxes of celestial bodies, the orbits of visual double stars, etc.

In this chapter we shall present the principles of photographic astrometry and their application to the problem of determining positions and proper motions of a large number of stars since photographic methods at present are widely used for the construction of star catalogs. It is our aim to give only general ideas without excessive details and to point out the difficulties encountered as we go along.

67. Astrograph and Its Structure. The Field of Photographic Images

The basic instrument used in astrometry for obtaining photographic observations is the *astrograph* (Fig. 60); it is used to photograph separate portions of the sky. The astrograph consists of a photographic tube on a parallactic mounting and a clock drive as well as a visual tube for guiding. The photographic tube or the camera consists of the objective, the tube proper, and a plateholder.

Fig. 60. Standard astrograph.

The objective of the astrograph forms an image of a portion of the celestial sphere on the photographic plate placed in its focal plane. The objective is the primary part of the astrograph since it determines the quality of the astrographic negative.

The tube must be rigid to avoid flexure, that is, any relative shift between the plateholder and the objective at various positions of the optical axis of the astrograph relative to the horizon.

The plateholder end resembles a nozzle, one end of which is inserted into the tube and the other end bears a flange for securing the photographic plate. The plateholder is set in its support and can easily be removed together with the photographic plate, which is protected by a special slide from exposure to light. The plateholder end can be moved along the optical axis of the astrograph and locked in any position. This is necessary in order to be able to place the photographic plate in the focal plane of the objective.

The photographic tube must have a shutter for making the exposure which is usually placed in front of the photographic plate (for short exposures, it is placed near the objective in order to achieve even illumination of the field). At the objective end of the tube a dewcap is placed in order to prevent the dew from gathering on the front surface of the objective lens.

The photographic tube is mounted on a parallactic mounting. The supporting column of the instrument, whose base is on a special foundation isolated from the dome, has a base plate and a bearing for the polar axis which is placed parallel to the earth's axis. Perpendicular to the polar axis is the declination axis. On one end of the declination axis is attached the photographic tube and on the other the counterweight. The objective end and the plateholder end of the tube are also balanced with respect to the declination axis.

When the instrument is rotated around the polar axis, the extension of the optical axis of the astrograph describes a diurnal parallel on the celestial sphere. By turning the instrument around the declination axis it is possible to shift the instrument to a diurnal parallel with a different declination. The graduated circles set on the axis make it possible to direct the instrument to any point on the celestial sphere according to its coordinates. The tube can be locked and then moved micrometrically relative to each axis by means of handles usually referred to as keys. Modern instruments are sometimes equipped with auxiliary electric motors by means of which the micrometric motion is more convenient than the manual drive.

The clock drive rotates the astrograph tube around the polar axis with a velocity equal to the rotation of the celestial sphere; thus, during the exposure, the instrument retains its line of sight toward the sky region photographed. In order to control the regularity of the instrument rotation by the clock drive, a visual telescope is used. By looking through this telescope, it is possible to correct the astrograph position. This visual tube, called the guiding telescope, is placed parallel to the photographic tube and for the purpose of guiding accuracy has the same focal length as the astrograph.

The image of the section of the celestial sphere formed on the photographic plate by the astrograph objective is a central projection of all the points in the region to be photographed from the secondary principal point of the objective to the photographic plate.

Let O_1 and O_2 be, respectively, the primary and the secondary principal points of the objective (in Fig. 61, we shall assume that they coincide). We construct the celestial sphere with its center at this point of

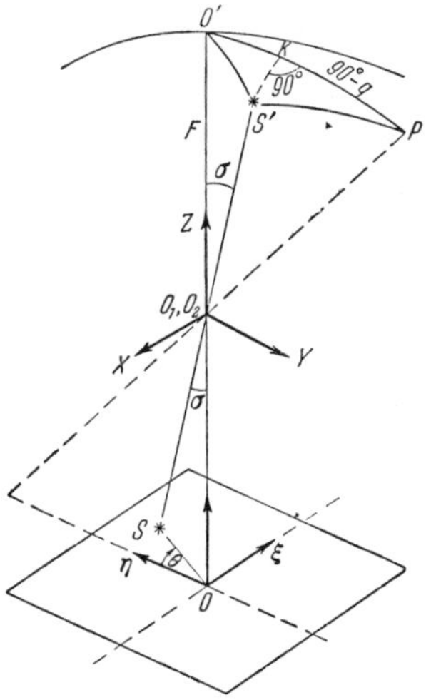

Fig. 61. Central projection of the celestial sphere on the plate.

the objective, with a radius equal to the focal length F. Let the star be on the celestial sphere at the point S'. Its image will be formed by the objective at the point S on the photographic plate. The base point, O, of the perpendicular dropped from the secondary principal point O_2 on the photographic plate is called the *optical center* of the plate. The distance between the optical center and the point where the plate is crossed by the optical axis defines the inclination of the photographic plate. The plate must be perpendicular to the optical axis, so that the image quality is not made worse due to the defects of the object lens. We assume that the photographic plate is placed so that $O_2O = F$.

Let the point O' on the celestial sphere correspond to the optical center of the plate. It is easy to derive the relation between σ, the angular distance from the star to the point O', and s, the linear distance between the star image and the optical center O:

$$s = F\tan\sigma. \qquad (42)$$

This relation is called the *tangent law*. It should be pointed out that the central projection of the great circle of the celestial sphere forms a straight line on the photographic plate.

The relation (42) determines the scale of the astrograph in the plane of the photographic plate. *The scale of the astrograph* m'' is the number of angular seconds corresponding to one millimeter on the photographic plate. It can be obtained easily if one puts s = 1 mm; then

$$m'' \approx \frac{1}{F} 206\,265''.$$

Let us set $s = a/2$, where a is the length of one side of the plate; then the area covered by the plate in units of degrees is as follows:

$$2\sigma \approx 57^\circ.3 \frac{a}{F}.$$

For example, in case of the wide-angle astrograph of the Moscow observatory, which has a focal length $F = 234$ cm and a photographic plate 24 × 24 cm, the quantity $m'' = 90''$ and the field covered is 6° × 6°.

Astrographs are divided into long-focus types (which have a large scale, about 30″ to 60″ per mm, and a small field, about 2° × 2°) and the short-focus or wide-angle type (which has a smaller scale, about 100″ per mm, but covers a larger area on the celestial sphere, 5° × 5° to 10° × 10°). As a rule, all long-focus astrographs have a smaller ratio of aperture to focal length, 1:20 to 1:15, and consequently less speed.

The rule $s = F \tan \sigma$ is only approximately fulfilled in practice. The distortions of the objective, the effect of the differential refraction and of differential aberration, as well as atmospheric dispersion, upset this relationship in the plane of the photographic plate.

68. Defects of the Objective and Distortions of the Photographic Image Field

It is extremely important in astrophotography to obtain on a smooth plate star images in the form of small circles which are sharply defined and are symmetrical in form and intensity. It is also important that the photographic image field be undistorted. Therefore the photographic objective must be subject to rigid requirements in order to be able to correct its faults. Here are the important defects: spherical aberration and coma, distortion, astigmatism and curvature of the field, chromatic aberration and prismatic behavior of the objective.

a) *Spherical aberration and coma.* Spherical aberration occurs when the rays of the light beam, parallel to the optical axis and converging toward the focus after passing through different annular zones of the object lens, are brought together at different points along the optical axis. This causes a symmetrically blurred image. Spherical aberration of slanted beams forming an image near the edge of the plate is called coma. In addition to blurring the image, the coma disturbs the image symmetry in form as well as in intensity of blackening. The image is elongated in the direction of the plate center. Due to coma, the center of the blackened image does not coincide with the central ray of the beam.

b) *Distortion* is the departure from the central projection of the image field on the photographic plate. Due to distortion, segments s on the plate do not conform to the tangent law but may be represented by an expression

$$s = F \tan \sigma + \nu_1 \tan^3 \sigma + \nu_2 \tan^5 \sigma + \ldots,$$

where the quantities ν_1 and ν_2 are called the distortion coefficients. A simultaneous correction of spherical aberration and distortion is possible only for long-focus objectives with a small field. In increasing the size of the field in high-speed astrographs it becomes preferable to have sharp images even if they are shifted by distortion; it is simpler to determine the distortion by special investigations and by compensating for it rather than to have the effect of asymmetrical and blurred images due to spherical aberration and coma.

c) *Astigmatism and curvature of the field.* Astigmatism appears as the result of oblique bundles of rays, and is caused by different foci for bundles of rays directed along two mutually perpendicular diameters. The circle of least scattering of the beam covering the entire objective occurs somewhere in the gap between the focus for radial beams and the focus for transverse beams. Circles of least scattering for various slanted beams are not in the plane perpendicular to the optical axis, but on some surface, referred to as the focal surface, which causes curvature of the image field. Consequently, it is impossible to obtain the smallest image diameters over the whole plate at once. The plate is usually placed at some optimum position so that the images at its center and near the edges are slightly blurred.

d) *Chromatic aberration* is subdivided into aberration of position and aberration of magnification. Due to the first, foci of rays with different wavelengths have different positions; this causes blurred images similar to spherical aberration. In aberration of magnification, the positions of the foci for oblique rays of various wavelengths are in the same focal plane, but at various places. Thus the focal lengths of rays of different wavelength are different, and the scales for various rays are therefore unequal. This second type of aberration is less tolerable since it transforms the star images into spectra elongated toward the plate center. Photographic objectives are usually designed so that rays for which the photographic emulsion is most sensitive—that is, blue and violet—converge at the focus in the best possible way.

e) *Prismatic effect of the objective* breaks up the star image into a spectrum; that is, the objective in this case has the characteristic of a prism with a small refraction angle. The prismatic effect of the objective frequently occurs when the lenses are poorly centered. This effect is studied by obtaining photographs of some sky region with two different positions of the telescope tube relative to the supporting column in such a manner that the prismatic effect of the objective works in opposite directions.

Correction and reduction of the aberrations is achieved by increasing the number of lenses of the objective, varying the kind of glass the lenses are made of, and properly selecting the radii of curvature, the thickness of the lenses and the distances between them. Aberrations are least near the optical axis. Consequently, the objectives of long-focus astrographs are simpler in construction and consist usually of two lenses. With an increase in the size of the field, the complexity of the lenses also increases.

Reflectors are free from chromatic aberration but have not, as yet, found general use for precise determination of positions due to the small field of view and the instability of the focal plane during temperature changes.

Let us now summarize what has been presented. Aberrations of the objective, besides blurring the images, decrease the limiting star brightness on the plate and also may cause a shift of the star images relative to each other. A purely geometric deformation of the whole image field is caused by distortion. A displacement of the brighter stars relative to the fainter stars—the magnitude equation—causes a systematic error, the elimination of which at present remains one of the serious problems facing photographic astrometry. Causes of the magnitude equation have been partially studied; among which may be mentioned: poor guiding, coma, astigmatism, elongation of the images into a spectrum. The latter arises because of chromatic aberration of the enlargement and the prismatic shape of the objective, and also from atmospheric dispersion. Elongation of images into spectra leads to a different shift of the center of darkening for stars of different spectral classes. An inclination of the plate may distort the aberration symmetry relative to the center of the plate.

Among the errors connected with external causes are differential refraction, atmospheric dispersion, and differential aberration. We shall not consider the scintillation of images since its influence is accidental and leads during long exposures only to some evenly distributed blurring of the image.

Refraction shifts the star images toward the zenith. If the refraction were even for the portion covered by the field of the photographic plate, then, despite the overall displacement, the relative positions of the stars would not change. But since refraction changes with zenith distance, the stars with zenith distances different from the one in the plate center will be displaced relative to the central stars. This phenomenon is called *differential refraction*. *Atmospheric dispersion*, causing elongation of the star images into spectra, particularly at great zenith distances, is the consequence of variation of refraction coefficient with the wavelength of the light. *Annual differential aberration* displaces the star images relative to the plate center in the same way as differential refraction.

Due to all these distortions the measured coordinates of stars on the plate will not satisfy the relations obtained on the basis of the tangent law.

69. Obtaining Astronegatives

Astrometric negatives are obtained on photographic plates. The glass, covered with a layer of light-sensitive emulsion, must be flat and not too thin so that it does not bend when placed in the plateholder and when it is measured on the measuring machine. If photographs are taken through the glass, it must be flat with plane-parallel surfaces and must be without cords or other blemishes. Close to the ideal is the so-called mirror glass.

The chromatic curve of the emulsion usually used in astrometry, with the maximum in the blue and violet region, basically corresponds to the chromatic curve of the photographic objective. The emulsion must have maximum sensitivity, but at the same time have a fine grain. These requirements, which are hardly compatible, are necessary to obtain images of faint objects and to guarantee that the position of the images will not depend upon the positions of separate light-sensitive emulsion grains. One should consequently point out that due to the photographic widening of images in the emulsion, the star image is formed not only by the grains on which the light acts, but also by the neighboring grains. In photographing bright objects, diapositive plates are used. The underside of the glass plate may be covered with a halo-reducing layer (usually a dye which dissolves during the developing of the plate). This is done for the purpose of decreasing the amount of light reflected back to the emulsion after passing through the glass, in this way, decreasing the light scatter causing the haloes around the star images.

The focusing of the astrograph—finding the optimum position of the plateholder for the sharpest images over the field—is made by obtaining several test exposures of some star region on one plate at different positions of the plateholder. The position of the focus depends upon the temperature, especially for long-focus astrographs, since the lens curvature and the tube length vary with the change of temperature. Therefore it is necessary to investigate this variation empirically and to set the plateholder according to the surrounding temperature.

Sky regions are usually photographed near the meridian so that the differential refraction will vary as little as possible with the change in elevation of the center of the region during the time of exposure. During this whole time, with the clock drive in operation, the observer makes certain that the "guide" star does not leave the intersection of the threads in the guiding eyepiece and corrects the position of the astrograph when necessary with the micrometric keys. A recent development is the use of automatic systems for control of the astrographs. This is done with the application of phototubes for automatic photoelectric guidance which corrects the motion of the instrument without the attention of the observer. The design principle in photoelectric guidance is similar to that for the photoelectric microscope of the automatic measuring machine described in Section 70.

At the end of an exposure, the data of the observation are written down on the emulsion with a pencil at the edge of the plate. The pertinent data are: the time of observation, the sky region, the instrument position (the tube on the east or west side of the column) and the observer's name. The plates are developed in darkness or with a weak red light in a developer designed for underexposed negatives and providing small grain. This is done because the images of faint objects, in which the astronomers are most interested, are always underexposed. The plates are fixed in the usual way. The use of hardening agents for strengthening the layer is not recommended because of possible deformations of the emulsion. The drying of the astronegatives must be even and not too rapid for the same reasons.

The astronegative obtained is in itself of irreplaceable value, and therefore must be carefully preserved. The astronegatives are kept in a dry and cool place, in envelopes or boxes with interlayers. One must use the astronegatives with great care and cleanliness, and avoid touching or marking the emulsion.

70. Measuring Astronegatives

Astronegatives are measured with special gauges which may be subdivided into screw and scale machines. Depending upon the construction of the gauge, one can measure either only one coordinate or both

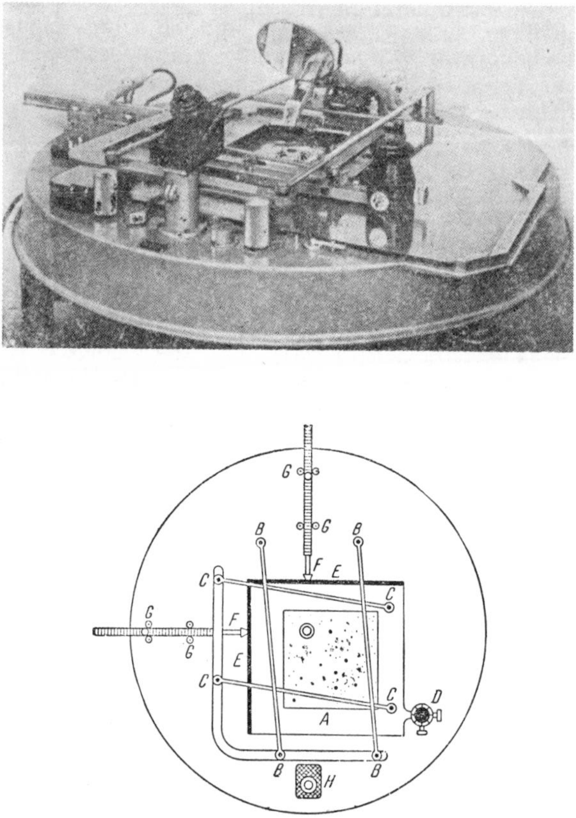

Fig. 62. General view and diagram of the coordinate measuring machine.

simultaneously. As an example, we give a short description of the KIM-3 coordinate measuring machine (Fig. 62).

On the top surface of the round base of the machine lies a large, flat glass plate on which the carriage of the table *A*, with the astronegative placed on it, can freely move on rollers. Because of the use of a system of two-hinged parallelograms *B* and *C*, the carriage with the plate may be moved only linearly, without rotation, parallel to the two mutually perpendicular directions. An approximate setting of the plate is made by moving the handle *D* connected with the carriage and resting freely on the glass plate. The fine displacements are accomplished with micrometer screws attached to this handle.

Two precisely constructed guides *E*, at right angles to each other, to which are pressed the end rollers of the scales *F*, are fastened to the carriage of the table. When the carriage with the astronegative is moved, the measuring scales are displaced along their own length perpendicular to the guides of the carriage on the rollers *G* fixed to the base of the machine. A permanent contact between the end rollers of the measuring scales and the guides of the carriage is made with springs.

The optical systems used for setting the images and for reading the scales are placed at the base of the machine and have an eyepiece H in common with all the microscopes. By switching the illumination from the scales to the plate through their respective light sources, it is posssible to see alternately either a section of the astronegative or one or the other of the two scales. Near the eyepiece are knobs for rotating the micrometers under the scales which are used for focusing and for switching the reversing prism. Measurement accuracy of the KIM-3 is approximately ±1.5 microns.

The general principles of astronegative measurement are briefly described. The plate is placed on the measuring machine with the emulsion facing the objective of the measuring microscope. The position of the plate when fastened to the platform must be identical with its position in the astrograph plateholder so that the conditions of the plate at the time of observation will be duplicated as closely as possible. On the glass side of the plate all objects to be measured, especially if there are not very many of them, can be encircled with ink and numbered, thus facilitating the process of finding them during measurement.

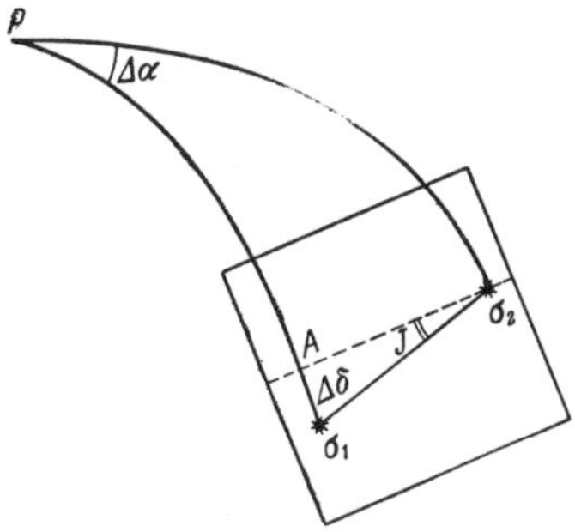

Fig. 63. Orientation of the plate.

The plate is oriented in the measuring machine so that one of the coordinates to be measured is, if possible, parallel to the declination circle passing through the optical center of the plate. The other direction to be measured is that perpendicular to it. This choice of coordinates to be measured facilitiates the subsequent analysis of the measurements. The orientation of the plate in the machine is made using two stars σ_1 and σ_2 with known coordinates (α_1, δ_1 and α_2, δ_2) near the edges of the plate. The plate is placed in the machine so that the straight line joining these two stars is parallel to the measuring screw or the machine scale. Then the platform together with the plate is rotated through an angle J, which is easily calculated from the triangle $\sigma_1 A \sigma_2$ (Fig. 63) according to the approximate formula:

$$\tan J = \frac{A\sigma_1}{A\sigma_2} = \frac{\Delta\delta}{\Delta\alpha \cos \frac{1}{2}(\delta_1 + \delta_2)}.$$

The settings on a star, depending upon the construction of the machine, are made either by moving the plate or the microscope, or by moving the threads of the microscope eyepiece. The thread is moved in the direction perpendicular to the coordinate to be measured; after measuring one of the coordinates of all the objects, the plate is rotated in the machine through 90°, then the second coordinate is measured. When both coordinates are measured simultaneously, settings are made with the crossthreads. For measuring faint objects, it is more convenient to use the double thread or the square formed by the two double threads. Settings on the graduations of the scale are made with the double thread or, as in the KIM-3 measuring machine, with a section of the double spiral of the helical micrometer. During measuring, all the final motions—shifting the plate and rotating the micrometer screw head—must be terminated with a motion in the same direction; that is, in the direction against the springs of the measuring machine.

During extended periods of measurement, it is necessary to keep a constant level of illumination of the astronegative (artificial illumination or natural light). At the beginning and at the end of each series of measurements one should carry out control settings on the same star images or on marks traced on the emulsion at the corners of the plate; thereby, in the case of a shift of the plate relative to the measuring machine, these displacements can be taken into account.

The measuring procedure introduces into the measured coordinates the following errors: instrumental, personal, and accidental. In order to account for the instrumental errors introduced by measuring with an imperfect measuring machine, the following error-introducing factors must be carefully investigated in advance: errors of the micrometer screws or scales; errors of the measuring microscope micrometer; departure from a straight line of the guides for shifting the platform with the plate and the measuring microscope; the angle between the plate and the optical axis of the microscope; the constancy of the measuring micrometer scale at different positions. The error corrections which affect the accuracy of the measurements are added to the measured coordinates.

The personal measurement error is part of the systematical error estimate of the center of the star image disk, distorting the readings in one way or in another for images of different diameters. This error is one of the causes of the magnitude error. In order to decrease this error, a reversing prism is placed in front of the measuring microscope eyepiece during measurement. By turning the prism through 90° (during this operation, the star image with the threads turns in the field of view through 180°), the settings on the image of each star are repeated and the mean of the coordinates measured at the two prism positions is taken. Since, after the reversing prism is rotated, the second settings on the disk are made from the same side as during the initial position of the reversing prism, there remains some doubt that a complete elimination of the personal error has been achieved. Therefore it is preferable to measure the plate again after the platform has been rotated through 180°.

Let D and R be the readings corrected for the instrumental errors and corresponding respectively to settings on the star image with the initial plate position and the plate position when rotated through 180°.

Let ΔX_D and ΔX_R be the personal measurement errors and X_0 the coordinate of the axis of rotation of the plate. Then the coordinates measured with respect to X_0 will be equal in both positions, respectively, to:

$$X_D = D + \Delta X_D - X_0, \quad X_R = X_0 - (R + \Delta X_R),$$

since $\Delta X_D = \Delta X_R$; then, taking the arithmetic mean, we obtain:

$$X = \frac{1}{2}(X_D + X_R) = \frac{1}{2}(D - R).$$

In such measurements, moreover, there appears the possibility of checking them, which is particularly important if the plate is measured during a period of several days. Subtracting the expression X_R from X_D, we obtain: $D + R - 2X_0 + 2\Delta X = 0$ or $\frac{D+R}{2} = X_0 - \Delta X$. Thus, for stars of the same magnitude, we must obtain equal values for the quantities $X_0 - \Delta X$, which enables one to check the stability of the measuring machine and to study the personal measuring errors.

Sometimes diffraction grids are used in front of the objective in order to decrease the magnitude error. Then supplementary fainter images are obtained which are distributed symmetrically around the central bright image. The arithmetic mean of the measured coordinates of the "satellite" images is used as the coordinates of the center of the bright image.

With the development of the photographic determination of coordinates of a large number of stars, automatic measuring machines are beginning to be accepted. Let us describe their principle of operation. The basis of this instrument is the screw-measuring machine whose microscope is equipped with a photoelectric attachment containing a photoelement. In front of the photoelement is a round diaphragm half of which is always covered. The diameter of the diaphragm is larger than the diameter of the star image. If the star image projected on the diaphragm surface does not coincide with the diaphragm center, then, upon rotating the diaphragm around its center, the light will strike the photoelement in pulses. The resulting photocurrent enters the guiding mechanism and, depending upon the phase and amplitude of the current, activates the motors. The motors rotate the screws which shift the microscope in one direction and the plate in the perpendicular direction. When the amplitude of the photocurrent becomes equal to zero, the screw rotation stops. Special counting devices connected with the screws record the measured coordinates on punch cards.

With a fully automatic measurement, the microscope is placed in accordance with the approximate rectangular coordinates recorded on the punch cards. The observer checks and corrects the microscope so that the measurement of the required star can be made. The mean square error of one setting with the automatic measuring machine is close to $\pm 0.5\mu$, which is three to four times smaller than the error made by visual measurement. Measuring with an automatic machine is several times faster than with ordinary measuring machines. The results of the measurements are recorded on punch cards together with the initial data and are fed into an electronic computer, which again simplifies and increases the speed of the analysis.

71. Ideal Coordinates and Their Interconnection with Equatorial Coordinates

From the measured coordinates x, y of objects photographed on the plate it is necessary to be able to calculate their equatorial coordinates α and δ. For this purpose, auxiliary rectangular coordinates ξ and η, called *ideal* or *standard* coordinates, are introduced. The establishment of the relation between the coordinates α, δ and ξ, η on the one hand and between ξ, η and x, y on the other makes it possible to solve in practice the problem of the determination of α, δ from the measured coordinates x, y.

The ideal coordinates were first introduced in practice by the Oxford astronomer Turner in 1893. The projection on the photographic plate of the declination circle crossing the point on the celestial sphere corresponding to the optical center was taken to be the η axis (see Fig. 61). The ξ axis is drawn perpendicular to the η axis. The axes are directed toward increasing declination and right ascension; the focal length of the objective is taken as the unit of scale of the ξ and η axes. Ideal coordinates for any point on the plate can be calculated from the known equatorial coordinates of this point and, conversely, equatorial coordinates may be calculated from known ideal coordinates. In deriving the formulas for the transformation of the coordinate systems, it is assumed that the spherical coordinates are transformed into coordinates on the tangent plane according to the laws of central projection. Distortions produced by the photographic method are taken care of during the transition from the ideal to the measured coordinates.

Let θ be the angle between the line joining the optical center with the star image S and the η axis; we then have

$$\xi = \tan\sigma \sin\theta, \quad \eta = \tan\sigma \cos\theta.$$

We designate the equatorial coordinates of the optical center for the accepted equinox by A_0, D_0. Then, by applying the basic group of spherical trigonometry formulas to the triangle $PO'S'$, we have:

$$\begin{aligned}
\cos\sigma &= \sin\delta \sin D_0 + \cos\delta \cos D_0 \cos(\alpha - A_0),\\
\sin\sigma \sin\theta &= \cos\delta \sin(\alpha - A_0),\\
\sin\sigma \cos\theta &= \sin\delta \cos D_0 - \cos\delta \sin D_0 \cos(\alpha - A_0).
\end{aligned}$$

Dividing the second and third equation, respectively, by the first, we obtain:

$$\left.\begin{aligned}
\xi &= \frac{\cos\delta \sin(\alpha - A_0)}{\sin\delta \sin D_0 + \cos\delta \cos D_0 \cos(\alpha - A_0)} = \\
&= \frac{\cot\delta \sin(\alpha - A_0)}{\sin D_0 + \cot\delta \cos D_0 \cos(\alpha - A_0)},\\
\eta &= \frac{\sin\delta \cos D_0 - \cos\delta \sin D_0 \cos(\alpha - A_0)}{\sin\delta \sin D_0 + \cos\delta \cos D_0 \cos(\alpha - A_0)} = \\
&= \frac{\cos D_0 - \cot\delta \sin D_0 \cos(\alpha - A_0)}{\sin D_0 + \cot\delta \cos D_0 \cos(\alpha - A_0)}.
\end{aligned}\right\} \qquad (43)$$

These expressions may be transformed by introducing an auxiliary quantity q, defined by the relation $\cot q = \cot\delta \cos(\alpha - A_0)$. The quantity q is interpreted geometrically as the declination of the base of the perpendicular dropped from the star to the declination circle which passes through the optical center (see Fig. 61). Then we obtain the formulas for computing ξ and η from the known quantities α, δ and A_0, D_0:

$$\xi = \frac{\cot q \tan(\alpha - A_0)}{\sin D_0 + \cos D_0 \cot q} = \frac{\cos q \tan(\alpha - A_0)}{\cos(q - D_0)},$$

$$\eta = \frac{\cos D_0 - \sin D_0 \cot q}{\sin D_0 + \cos D_0 \cot q} = \frac{\sin(q - D_0)}{\cos(q - D_0)} = \tan(q - D_0).$$

Now we shall derive the formulas for the reverse problem of determining α and δ from the known values of ξ, η and A_0, D_0. For this purpose we shall transform the equations (43):

$$\eta \sin D_0 + \eta \cot\delta \cos D_0 \cos(\alpha - A_0) =$$
$$= \cos D_0 - \cot\delta \sin D_0 \cos(\alpha - A_0),$$

$$\xi \sin D_0 + \xi \cot\delta \cos D_0 \cos(\alpha - A_0) = \cot\delta \sin(\alpha - A_0).$$

From the first line we easily get:

$$\cot\delta \cos(\alpha - A_0) = \frac{1 - \tan D_0}{\eta + \tan D_0}, \tag{44}$$

and from the second, by using this relation,

$$\cot\delta \sin(\alpha - A_0) = \frac{\xi \sec D_0}{\eta + \tan D_0}. \tag{45}$$

We find the right ascension by dividing (45) by (44) and by computing $(\alpha - A_0)$ according to the formula $\tan(\alpha - A_0) = \frac{\xi \sec D_0}{1 - \eta \tan D_0}$; the declination δ is then found from the formula (44) or (45), depending on the quantity $(\alpha - A_0)$.

All the formulas have been derived on the assumption that the equatorial coordinates of the optical center are known accurately. Let us consider the question of determining the quantities A_0 and D_0, the effect of errors in their determination, and the derivation of the relation between the ideal coordinates for plates with various optical centers.

72. Determination of the Position of the Optical Center. Photographic Plates with Different Optical Centers

In order to find the equatorial coordinates of the optical center, it is necessary to determine its position on the astronegative. There are several laboratory methods for determining the position of the optical center on the plate. The simplest method is the following, which is used with the majority of astrographs, particularly the ones with a greater focal length.

A diaphragm with an illuminated pinhole in the center is put over the astrograph objective. A transparent glass plate is inserted into the open plateholder attached to the astrograph. By holding in his hands a corner of a white sheet of paper at some distance from the plate and looking through the glass into the astrograph, the observer tries to superimpose the paper corner, its image formed by reflection from the glass plate, and the image of the diaphragm pinhole. This image of the pinhole is the optical center. Marking with ink the optical center and the corners of the plateholder on the glass, we can determine its position relative to the geometrical center of the plate. By knowing the equatorial coordinates of the standard stars and by measuring their distance from the geometrical center of the plate, it is then possible to determine the coordinates of the optical center A_0, D_0. The determination of the position of the optical center must be made for various altitudes and azimuth angles of the tube in order to find a possible dependence of the optical center position on the flexure of the astrograph tube.

In order to estimate the inclination of the plate to the optical axis of the objective it is necessary to find the point of intersection between the optical axis and the glass plate. For this purpose the observer, by holding in front of the plate a point source of light and by looking inside the tube, tries to place the light source on a straight line with its images reflected from the lens surfaces of the objective, first in one direction and then in a direction perpendicular to it.

By drawing ink lines each time on the plate through these images, we obtain an intersection point where the optical axis crosses the plate. The distance between the optical center and the intersection point characterizes the inclination of the plate to the optical axis.

The ideal coordinates for the same stars on two plates having different optical centers will not be the same. Let us determine this difference. We shall assume that the principal points of the objective coincide at the point O_2 (Fig. 61); for this derivation this is not essential. Let us introduce a space system of coordinates with the origin at O_2. The Z axis is directed along the $O'O_2$ line, and the X and Y axes are drawn parallel to the ideal coordinates. Let us measure the coordinates X, Y, Z in units of the focal length. If we draw a sphere with the radius F and designate the point of intersection between the straight line O_2S and the sphere by S'. the rectangular coordinates of this point will be (X, Y, Z), and the coordinates of the point S, that is, the central projection of the point S' on the plate, will be ξ, η.

Then we shall have: $O_2S = \sqrt{1+\xi^2+\eta^2}$, and the ratio $\frac{O_2S}{O_2S'} = \frac{\sqrt{1+\xi^2+\eta^2}}{1} = \frac{\xi}{X} = \frac{\eta}{Y} = \frac{1}{Z}$, from which we obtain:

$$X = \frac{\xi}{\sqrt{1+\xi^2+\eta^2}}, \qquad Y = \frac{\eta}{\sqrt{1+\xi^2+\eta^2}}, \qquad Z = \frac{1}{\sqrt{1+\xi^2+\eta^2}}. \tag{46}$$

Similarly for the plate with another optical center we have the coordinates of the point S in the second system of space coordinates:

$$X' = \frac{\xi'}{\sqrt{1+\xi'^2+\eta'^2}}, \quad Y' = \frac{\eta'}{\sqrt{1+\xi'^2+\eta'^2}}, \quad Z' = \frac{1}{\sqrt{1+\xi'^2+\eta'^2}}. \tag{47}$$

Transformation from one space system to the other with a common origin is made using the linear equations

$$\left.\begin{aligned} X' &= \alpha_1 X + \alpha_2 Y + \alpha_3 Z, \\ Y' &= \beta_1 X + \beta_2 Y + \beta_3 Z, \\ Z' &= \gamma_1 X + \gamma_2 Y + \gamma_3 Z, \end{aligned}\right\} \tag{48}$$

where the α_i, β_i, γ_i, are the direction cosines of the axes of one system with respect to the other.

Substituting expression (46) and (47) into (48), we obtain:

$$\begin{aligned} \frac{\xi'}{\sqrt{1+\xi'^2+\eta'^2}} &= \frac{\alpha_1\xi + \alpha_2\eta + \alpha_3}{\sqrt{1+\xi^2+\eta^2}}, \\ \frac{\eta'}{\sqrt{1+\xi'^2+\eta'^2}} &= \frac{\beta_1\xi + \beta_2\eta + \beta_3}{\sqrt{1+\xi^2+\eta^2}}, \\ \frac{1}{\sqrt{1+\xi'^2+\eta'^2}} &= \frac{\gamma_1\xi + \gamma_2\eta + \gamma_3}{\sqrt{1+\xi^2+\eta^2}} \end{aligned}$$

or, dividing the first two expressions by the last one, we obtain

$$\xi' = \frac{\alpha_1\xi + \alpha_2\eta + \alpha_3}{\gamma_1\xi + \gamma_2\eta + \gamma_3}, \quad \eta' = \frac{\beta_1\xi + \beta_2\eta + \beta_3}{\gamma_1\xi + \gamma_2\eta + \gamma_3}. \tag{49}$$

Consequently, the ideal coordinates of two plates having different optical centers are interrelated by a ratio of linear expressions.

In practice the errors in determining the position of the optical center are small since the cosines of the angles α_1, β_2, γ_3 between the corresponding axes are close to unity, and the remaining direction cosines are close to zero. Therefore, we set $\alpha_1 = \beta_2 = \gamma_3 = 1$ and expand the denominators in series. Thus we obtain for ξ':

$$\begin{aligned} \xi' &= (\xi + \alpha_2\eta + \alpha_3)(1 - \gamma_1\xi - \gamma_2\eta - \ldots) = \\ &= \xi + \alpha_2\eta + \alpha_3 - \gamma_1\xi^2 - \gamma_1\alpha_2\xi\eta - \alpha_3\gamma_1\xi - \gamma_2\eta\xi - \alpha_2\gamma_2\eta^2 - \\ &\quad - \alpha_3\gamma_2\eta + \ldots = (1 - \alpha_3\gamma_1)\xi + (\alpha_2 - \alpha_3\gamma_2)\eta + \\ &\quad + \alpha_3 - \gamma_1\xi^2 - \alpha_2\gamma_2\eta^2 - (\alpha_2\gamma_1 + \gamma_2)\xi\eta + \ldots \end{aligned}$$

We shall neglect the products of small cosines with η^2 and $\xi\eta$ and also take into account that

$$(1 - \alpha_3\gamma_1) = 1, \quad (\alpha_2 - \alpha_3\gamma_2) = \alpha_2, \quad \alpha_3 = -\gamma_1 \text{ and } \beta_3 = -\gamma_2.$$

We then obtain:

$$\xi' = \xi + \alpha_2\eta - \gamma_1 - \xi(\gamma_1\xi + \gamma_2\eta)$$

and analogously

$$\eta' = \eta - \alpha_2\xi - \gamma_2 - \eta(\gamma_1\xi + \gamma_2\eta).$$

From relations (49), and putting $\xi' = \eta' = 0$ and $\gamma_3 = 1$, we may conclude that $\gamma_1 = \xi_0$, $\gamma_2 = \eta_0$ are essentially the coordinates of the projected optical center of the second plate on the first plate, or that $\Delta A_0 \cos D_0$ and ΔD_0 may be considered as errors in the accepted values of A_0 and D_0.

Terms of the first order are not of great interest since they are calculated during analysis of the measurements, as will be shown further on. Thus, in order to simplify calculations, it is desirable to deal with the terms separately (beginning with second-order terms). Then the formulas for evaluating the effect of error in the coordinates of the optical center on the calculated ideal coordinates will have the form

$$\Delta\xi = \xi(\Delta A_0 \cos D_0 \xi + \Delta D_0 \eta),$$
$$\Delta\eta = \eta(\Delta A_0 \cos D_0 \xi + \Delta D_0 \eta).$$

Since ξ and η are small quantities, it follows from these formulas that the coordinates A_0 and D_0 of the optical center must be known in any case to an accuracy one order lower than the accepted accuracy of the ideal coordinates.

73. Relation between Measured and Ideal Coordinates

The ideal coordinates are a purely geometric construction and therefore the point on the plate defined by the coordinates ξ and η does not coincide with the image center of the star, which is displaced for a number of reasons. The axes of the measured coordinates x, y also do not coincide with the coordinate axes ξ, η. Consequently the ideal coordinates ξ, η are not identical with the measured coordinates x, y, although they differ from them only slightly. Turner represented the relation between these two systems as a series in terms of the measured coordinates x, y:

$$\xi - x = ax + by + c + a'x^2 + b'xy + c'y^2 + \ldots,$$
$$\eta - y = dx + ey + f + d'x^2 + e'xy + f'y^2 + \ldots,$$

where the coefficients a, b, c, d, e, f, and a', b', c', d', e', f' are constant quantities called *plate constants* or *coefficients of the Turner equations*.

We note that to be comparable with the measured coordinates, ξ and η must be multiplied by the quantity F, since ξ and η are calculated in units of focal length. The inaccuracy in the quantity F, as will be seen further on, is accounted for in the coefficients of the first-order terms. In the majority of cases, when the measured region does not encompass many degrees on the celestial sphere, it is sufficient to limit oneself to the first-order terms in the Turner equations.

The validity of the application of the Turner equations may be proven by analyzing the causes that contribute to the discrepancy between the ideal and the measured coordinates. Let us consider these causes separately.

a) The non-coincidence of the ideal and measured coordinate systems is caused by the following: the origins of the two systems do not coincide, the axes of the measured coordinates are turned relative to the ideal axes, their scales are different and the angle between the measured axes is not equal to 90°. These geometrical differences between the two systems are the consequence of the errors in the orientation of the plate in the measuring machine and the errors of the measuring machine itself. The transformation of the coordinates x, y into the coordinates ξ, η, with these geometrical differences of the systems, is an affine transformation and, according to analytical geometry, is represented by a linear form

$$\xi - x = a_1 x + b_1 y + c_1,$$
$$\eta - y = d_1 x + e_1 y + f_1.$$

b) The measured coordinates must be referred to the optical center with respect to which the ideal coordinates are calculated. The ideal coordinates are calculated with the use of the assumed coordinates A and D of the optical center. But in reality the optical center differs from the point assumed to be the optical center by the quantities ΔA and ΔD; thus arises the discrepancy between the calculated ideal coordinates and the coordinates measured on the plate. The same effect appears if one takes as the optical center the point of intersection of the plate and the optical axis of the objective; then this discrepancy is caused by the inclination of the plate.

Let ξ' and η' be the ideal coordinates computed with the assumed coordinates A_0 and D_0 of the optical center, let ξ and η be the values $A_0 + \Delta A_0$ and $D_0 + \Delta D_0$ respectively, and let x and y be the measured coordinates of the real plate. Then, assuming that the origin of the (x, y) system must lie at the true optical center and that the axes x, y are oriented along ξ, η, one may write the relations

$$\left.\begin{aligned} \xi &= x + \Delta A_0 \cos D_0, \\ \eta &= y + \Delta D_0. \end{aligned}\right\} \qquad (50)$$

The relation between the calculated ideal coordinates ξ', η' and the measured coordinates x, y may be obtained by using equations (50) and neglecting the second order terms:

$$\xi' - x = \alpha_2 y - x(\Delta A_0 \cos D_0 x + \Delta D_0 y),$$
$$\eta' - y = -\alpha_2 x - y(\Delta A_0 \cos D_0 x + \Delta D_0 y).$$

The correction of the measured coordinates will be expressed by a formula with linear and quadratic terms, as was shown in Section 72 for plates with different optical centers. Consequently,

$$\xi - x = a_2 x + b_2 y + c_2 + a_2' x^2 + b_2' xy + \ldots,$$
$$\eta - y = d_2 x + e_2 y + f_2 + e_2' xy + f_2' y^2 + \ldots,$$

The coefficients for third-order terms are in all cases small and can be ignored. Since the accepted coordinates of the optical center are practically close to their true values, the second-order terms can also be frequently neglected.

c) The relative positions of the star images on the plate are distorted due to differential refraction and differential aberration. Refraction and aberration shift the images of various stars on the plate by different amounts depending on their distance from the zenith and the apex.

As can be seen from rather cumbersome expressions obtained by using the formulas of spherical astronomy, the influence of these effects can be represented by the following relations:

for differential refraction
$$\xi - x = a_3 x + b_3 y + c_3 + a_3' x^2 + b_3' xy + c_3' y^2,$$
$$\eta - y = d_3 x + e_3 y + f_3 + d_3' x^2 + e_3' xy + f_3' y^2;$$

for differential aberration
$$\xi - x = a_4 x + b_4 y + c_4 + a_4' x^2 + b_4' xy + c_4' y^2,$$
$$\eta - y = d_4 x + e_4 y + f_4 + d_4' x^2 + e_4' xy + f_4' y^2.$$

Thus the difference between the ideal and measured coordinates, regardless of its causes, may be represented by a power series.

74. Determination of the Relative Equatorial Coordinates of Stars

Since an outline of the photographic work for constructing star catalogs is to be given in later chapters, we shall consider here only the problem of determining the coordinates for all stars of the region photographed on the plate. The numerous methods for the determination of coordinates of individual objects will not be considered, with the exception of the Schlesinger method for determining positions of minor planets.

To obtain a catalog of star positions for a large part or zone of the celestial sphere the photography is conducted so that consecutive photographs in a sequence will overlap; that is, the centers of the plates of each series will fall in with the corners of the succeeding series of plates. This is done in order to decrease somewhat the field errors when the arithmetic mean of the coordinates obtained by two plates is formed.

In order to determine the relative equatorial coordinates it is necessary to know the equatorial coordinates and the proper motions in the system of the fundamental catalog for some of the stars (the reference stars) on the plate. Usually 15 to 25 reference stars are chosen from the plate. The reference stars within the region must be evenly distributed. This is important for a more certain determination of the plate constants. If the reference stars are evenly distributed, it is easier to take into account the field distortions.

The equatorial coordinates α_j, δ_j of the reference stars are referred to some equinox with respect to which the coordinates of the stars are to be determined. The proper motions μ_α and μ_δ are taken into account for the time interval between the catalog equinox and the epoch in which the astronegative is obtained. The coordinates α_j and δ_j are calculated according to the formulas:

$$\alpha_j = \alpha_k + P_\alpha (t - t_k) + \mu_\alpha (t_0 - t_k),$$
$$\delta_j = \delta_k + P_\delta (t - t_k) + \mu_\delta (t_0 - t_k),$$

where t is the equinox used for the analysis, t_k is the equinox of the reference catalog, t_0 is the epoch of the photograph, P_α and P_δ are the precession in α and in δ. From the coordinates α_j, δ_j, and the accepted coordinates A_0, D_0 of the optical center, the coordinates ξ_j, η_j of the reference stars with respect to the accepted equinox are computed.

By measuring the images of all stars on the plate, the measured coordinates x_j, y_j of the reference stars and x_i, y_i of the stars to be determined are obtained. These coordinates are then corrected for all known measuring errors. After this the Turner equations for the reference stars are written:

$$\left.\begin{aligned} \xi_j &= ax_j + by_j + c, \\ \eta_j &= dx_j + ey_j + f \end{aligned} \quad (j = 1, 2, \ldots, n). \right\} \tag{51}$$

These formulas are usually limited to linear terms if the region photographed is not too near the horizon or if the measured portion of the plate is small and the coordinates of the optical center are sufficiently defined. For plates with a large field, it is sometimes necessary to introduce into the Turner equations second-order terms or to take them into account separately by using them to correct the measured coordinates. The latter is done for second-order terms of the differential refraction and aberration. In case distortion is evident, it is investigated and also taken into account.

By solving the conditional equations (51) by the method of least squares we obtain the plate constants a, b, c, and d, e, f. Substituting these values back into equations (51), we can obtain the residuals of the equations. A large residual indicates an error in the measurement or an inaccuracy in the coordinates of a reference star. Reference stars with large residuals must be excluded from the system, after which new plate constants are obtained.

Now it is possible to calculate for the observed stars their ideal coordinates ξ_i, η_i, by using the derived plate constants and the measured coordinates x_i, y_i, from the relations

$$\xi_i = ax_i + by_i + c,$$
$$\eta_i = dx_i + ey_i + f.$$

To obtain α_i, δ_i, the equatorial coordinates of the observed stars, we need only make the transition from ξ_i, η_i according to the formulas

$$\cot \delta_i \sin(\alpha_i - A) = \frac{\xi_i \sec D}{\eta_i + \tan D},$$
$$\cot \delta_i \cos(\alpha_i - A) = \frac{1 - \eta_i \tan D}{\eta_i + \tan D}.$$

The derived coordinates α_i, δ_i are in a system which does not exactly coincide with the system of coordinates of the reference stars because of the influence of the systematic errors involved in the photographic method of determining coordinates. In many cases systematic differences between photographic catalogs are rather large.

Coordinates of close binary stars cannot be obtained to sufficient accuracy by the photographic method due to photographic effects which change distance between the centers of the two star images. Among them is the so-called Kostinsky effect. This causes a repulsion between the components of close double stars and is the result of exhaustion of the developer at the portion between the two star images. Coordinates of close double stars must therefore be determined visually.

75. Determination of the Relative Proper Motions of Stars

To determine the proper motions of stars, it is necessary to have at least two astronegatives in the region of investigation, with a sufficiently large interval of time between the two observations. The adequacy of the interval between epochs is generally determined by the scale of the plate. It is computed that for an astrograph with a focal length of about five meters, the proper motions can be determined with a time interval exceeding twenty-five years. This guarantees an accuracy of $\pm 0''.003$ of the derived proper motions.

It is necessary to obtain the plates of the second epoch with the same instrument used for the first epoch since photographic as well as visual observations have significant systematic errors depending partly upon the instrument. In obtaining plates of the second epoch it is necessary to conform as far as possible to the conditions of the first epoch, obtaining image shapes close to those on plates of the first epoch. This simplifies the calculations and eliminates from the results some possible systematic errors.

The plates of the second epochs must be exposed with the same instrument position relative to the column and having an emulsion with a color sensitivity close to that used for the first observations. The observation data and the hour angle at the middle of the exposure must be close to those for the plates of the first epoch. The plates must have similar optical centers; that is, the guiding must be done with the same star, the same threads, and the same plateholder (the latter assures a similar plate inclination).

The plates must be measured by the same method. In fastening the plates on the measuring machine, one must place them in the same positions with respect to the measuring scales or screws. Since the differences in coordinates of stars on the two plates will later be studied, the systematic errors of measured differences will be significantly reduced. The orientation of the plates is made using stars with known α and δ, so that the differences of the measured coordinates, Δx and Δy, will correspond to $\Delta\alpha \cos \delta$ and $\Delta\delta$, and that it will not be necessary to take into account the inclination of the axes of the measured coordinates relative to the projection of the network of celestial coordinates.

Fig. 64. Blink comparator.

Special, so-called blink-comparator (Fig. 64) measuring machines have been constructed which make it possible to measure simultaneously the difference between the rectangular coordinates on the two plates.

For this purpose both plates are placed side by side on the moving carriage of the comparator and are oriented approximately the same way. The images are observed with two microscopes with their objectives placed above the corresponding sections of the plates, the latter being illuminated from underneath. The microscopes have a common eyepiece through which one may see both images either at the same time or separately on each plate. The differences Δx and Δy are measured with the eyepiece of the double micrometer. It is important that the relative position of the plates during measurement remain unchanged. This is an essential requirement for the comparator. The advantage of this instrument is that the settings on the two corresponding images are made almost simultaneously and, consequently, on the same type of images with similar personal errors. In measuring blurred or asymmetrical objects (extragalactic nebulae) it is particularly important to make the settings on image centers chosen in the same way. When the plates are measured separately, a change in the personal error may occur during the interval between the two measurements, and hence the personal error will not be completely eliminated from the coordinate differences. For these reasons the magnitude error is very much reduced in proper motions measured by the differential method.

The second epoch is sometimes photographed through glass; that is, on a plate set in the plateholder with the emulsion away from the objective (the anti-halation layer naturally must be removed). It is then possible, by placing the two astronegatives together with their emulsions facing each other, to measure the coordinate differences with ordinary measuring machines. Blemishes in the glass may cause errors while taking the photographs and while measuring through the glass; thus this method has not found widespread application.

We may write down the relations for the differences between the coordinates measured on the two plates. These relations are the basis for the determination of proper motions. We shall assume that the coordinates x, y for both plates differ from each other basically for the same reasons that cause the differences between the x, y and ξ, η coordinates on one plate. If the systems for measured coordinates for both plates are sufficiently close to each other, then

$$\left.\begin{aligned} \Delta x &= ax + by + c + \mu_x (t_2 - t_1), \\ \Delta y &= dx + ey + f + \mu_y (t_2 - t_1), \end{aligned}\right\} \tag{52}$$

where μ_x and μ_y are the proper motions along the x and y axes, and $(t_2 - t_1)$ is the difference between the plate epochs.

Since the quantities Δx and Δy are small, the rectangular coordinates x and y need not be precisely determined.

These relations are true if the conditions under which the first epoch observations were made are followed with sufficient accuracy for the second epoch observations. Otherwise there will appear systematic errors which cannot be accounted for. Thus a difference between the optical centers and a different hour angle corresponding to mid-exposure, which will change the amount of differential refraction, make it necessary to include second-order terms. If the calculations are only slightly complicated by violating these conditions, the possible displacements of the images of individual stars, which cannot be accounted for, are especially troublesome. These are caused by deviations of instrument positions relative to the column and by large differences in the time of observation. The first, due to the prismatic effect of the objective, causes a shift of the image center, depending upon the spectral class; the second causes a displacement due to the annual parallax.

Thus the problem of determining the proper motions is reduced to determining the coefficients a, b, c, d, e, f. After this, since the system of measured coordinates is close to the ideal system, μ_α and μ_δ are found from the formulas:

$$\mu_\alpha \cos\delta = \frac{M\Delta x'}{t_2 - t_1},$$

$$\mu_\delta = \frac{M\Delta y'}{t_2 - t_1},$$

where $\Delta x' = \Delta x - ax - by - c$, $\Delta y' = \Delta y - dx - ey - f$, and M is the plate scale.

If the coefficients a, b, c, d, e, f are determined from the solutions of conditional equations set up for stars with known proper motions, then the proper motions for all the remaining stars can be considered absolute, or rather, in the system of the accepted proper motions of reference stars.

However, in practice it is difficult to choose a sufficient number of reference stars with well-known proper motions. Therefore, reference stars are chosen on the plate according to their distances, that is, fainter ones, on the assumption that they have small proper motions. Then, equations (52) are solved on the assumption that for the reference stars μ_x and μ_y are equal to zero. Since this is not true in general, proper motions μ_x and μ_y, after calculation by the determined coefficients a, b, c, d, e, f, will relate only to some chosen group of reference stars with some kind of motion; that is, without taking into account the mean proper motions $\overline{\mu_x^0}$ and $\overline{\mu_y^0}$ of the reference stars. In fact, what can be done to reduce the effect of μ_x and μ_y is to throw out reference stars which, in solving the Turner equations, show large residuals. These large residuals are evidence of large proper motions or measurement errors.

If some of the stars on the plate have known proper motions μ_x' and μ_y', in the system of the fundamental catalog, then the determined quantities μ_x and μ_y can be reduced to this system. For this purpose we add the quantities $\overline{\mu_x' - \mu_x}$ and $\overline{\mu_y' - \mu_y}$ to all μ_x and μ_y values. These values $\overline{\mu_x' - \mu_x}$ and $\overline{\mu_y' - \mu_y}$ are the mean deviations between the relative and the catalog motions, which can be assumed to be equal to $\overline{\mu_x^0}$ and $\overline{\mu_y^0}$, the mean motion of the reference stars.

The method of referring the proper motions of stars to the extragalactic nebulae (Section 82) is the best method to set the system of proper motions on an absolute basis, but it is not always applicable to individual parts of the sky and is not convenient for all astrographs. Other methods for setting the system on an absolute basis founded on statistical laws are not good enough from the astrometric point of view. They are used in stellar astronomy and will not be presented here (those interested are referred to any course in stellar astronomy).

CHAPTER XI

The Fundamental System of Positions and Proper Motions of Stars; Its Derivation and Methods of Its Improvement

76. Catalogs of Stellar Positions and Proper Motions

Observed star coordinates are incorporated into catalogs. A catalog of positions is a list of a number of stars which contains results based on observational data. These results must be sufficient to determine the position of the mean equatorial coordinate system either for any epoch or for some specific one. The star catalogs may be divided into two groups: initial catalogs, which contain the results of direct observations, and derived catalogs, which are obtained by combining groups of initial catalogs.

The initial catalogs are further divided into absolute catalogs (obtained on the basis of absolute observations) and relative catalogs (obtained by a differential method). The Pulkovo catalogs of bright stars can serve as examples of absolute catalogs for which the coordinates were obtained regularly since the foundation of the observatory. In compilation of relative catalogs one usually aims first at determining the coordinates of a large number of stars within some fundamental system. As an example of the relative catalogs we may take the zone catalogs of the "Astronomische Gesellschaft" for all stars down to 9.0 magnitude, as well as the catalogs of photographic re-observation of these stars.

The initial catalogs contain the coordinates of the observed stars with respect to some equinox and the epoch of the observation of each star.

One should distinguish between the equinox of the initial catalog and the epoch of observation of the coordinates. The latter is the mean time of observation of a given star contained in the catalog. The apparent coordinates derived from observations give the position of the star on the celestial sphere at the time the observations were made;

these are reduced to the beginning of some year, for example, 1875.0, 1900.0, 1925.0, 1950.0, which is known as the equinox of the catalog. Thus, the star coordinates corresponding to the position at the time of observation are given with respect to a coordinate network related to a different time, that is, to the equinox epoch. This is done for the convenience of comparing the coordinates of a star from different initial catalogs in investigating the latter and for the derivation of the proper motion of stars.

For the purpose of making the reductions easier, the annual precession and its hundred-year variation are given for each star, while for stars with high declinations the third term of the precession is kept. For the purpose of identifying stars, their magnitude, their number from another catalog (usually according to the "Bonner Durchmusterung") and other particular characteristics of the stars are given, in addition to their serial number (the stars are arranged in order of increasing right ascensions). Besides the mean positions, separate observations are sometimes included in the initial catalog together with the data of each observation, the instrument position, the name of the observer and other data required for derivation of the mean positions of the stars in the catalog. Initial catalogs serve as material for obtaining derived catalogs which have greater accuracy due to the combination of several catalogs.

The derived catalogs may be fundamental or combined. By means of either of these, it is possible to construct a system of the mean equatorial coordinates for any epoch. The distinction between these catalogs consists in the fact that in constructing a fundamental catalog a new and original fundamental system is derived, while the combined catalog is obtained in an already known system. Fundamental catalogs as well as combined catalogs contain all the data necessary for the derivation of mean coordinates of the stars listed for any epoch, i.e., the mean coordinates of the stars reduced to the equinox of the catalog and with consideration of the proper motions as well; the yearly and the centennial proper motions of stars; and the precession terms. The annual precession and proper motion are frequently combined, forming the annual variation V. A. (Variato Annua), which makes it easier to take them into account.

The fundamental catalogs are obtained by combining the absolute and relative catalogs for different epochs. Absolute catalogs are used in order to obtain the new and improved fundamental system of coordinates and proper motions. In order to improve the catalogs with respect to accidental errors in coordinates and in proper motions of stars which determine the fundamental system, supplementary relative catalogs are brought in. The third fundamental catalog, FK3, whose system is accepted as a basis for all astronomical almanacs, can serve as an example of a fundamental catalog.

A combined catalog is made by incorporating the relative catalogs from several observatories for one and the same list of stars and for approximately the same epoch of observation in order to increase the accuracy of the star coordinates with respect to accidental errors and in order to decrease the influence of systematic errors of individual instruments. Proper motions, derived by using various other sources, are usually given in combined catalogs. The Catalog of Geodetic

Stars, KGZ, compiled from observations at five Soviet observatories is an example of a combined catalog.

Catalogs of proper motions are obtained by comparing several initial catalogs of positions for different epochs, or from especially set up photographic observations. Catalogs of proper motions are divided into absolute catalogs obtained from meridian instrument observations and relative catalogs obtained from photographic observations. An example of absolute proper motions are the proper motions derived in forming a fundamental system. The Pulkovo catalog of proper motions in the Kapteyn Selected Areas is an example of a catalog of relative proper motions.

77. Systematic Errors and Systematic Differences Between Catalogs

The primary characteristic of a catalog is the quality of the coordinate system which it establishes on the celestial sphere. Because of the imperfections of the instruments and the errors in the observations, the coordinates of stars in the catalogs contain accidental as well as systematic errors.

The error that distorts the coordinates of all stars in the same way within a certain region of the celestial sphere is called *the systematic error* of the catalog. This changes smoothly from region to region and is a function of α and δ. On the basis of careful and thorough investigations of numerous catalogs, systematic errors of right ascensions and declinations can be represented in the following form:

$$\Delta\alpha = \Delta A + \Delta\alpha_\alpha + \Delta\alpha_\delta + \Delta\alpha_m,$$
$$\Delta\delta = \Delta\delta_\alpha + \Delta\delta_\delta.$$

The subscripts designate the argument upon which one or the other component of a systematic error depends. For detailed investigations it is assumed that the quantities $\Delta\alpha_\alpha$ and $\Delta\delta_\alpha$ also depend upon declination and they are determined for various declination zones separately. Systematic errors of the form $\Delta\alpha_\alpha$, $\Delta\alpha_\delta$, $\Delta\delta_\alpha$ and $\Delta\delta_\delta$ distort the coordinate network. The $\Delta\alpha_\delta$ errors distort the declination circles and the $\Delta\alpha_\alpha$ errors affect their uniformity. Similarly the errors of the type $\Delta\delta_\alpha$ and $\Delta\delta_\delta$ affect the diurnal parallels (Fig. 65). The already known equinox correction ΔA causes a rotation of the coordinate network as a whole around the earth's axis. Sometimes the equator correction ΔD is determined.

The most outstanding systematic errors are of the type $\Delta\alpha_\delta$ and $\Delta\delta_\delta$, which are primarily caused by instrumental errors not taken into account. In connection with $\Delta\alpha_\delta$, one should point out the pivot irregularities, the side flexure, the lateral refraction, the changes in collimation, and personal errors depending upon the motion of the star. The $\Delta\delta_\delta$ error is caused by the instrument flexure, unaccounted graduation errors of the circle, as well as by the anomalous refraction. Systematic errors of the form $\Delta\alpha_\alpha$ and $\Delta\delta_\alpha$ are considerably smaller than $\Delta\alpha_\delta$ and $\Delta\delta_\delta$; they are the consequence of the changes of the external observation conditions from season to season and from the evening to the morning. The determining factor is the temperature variation, and it is not surprising

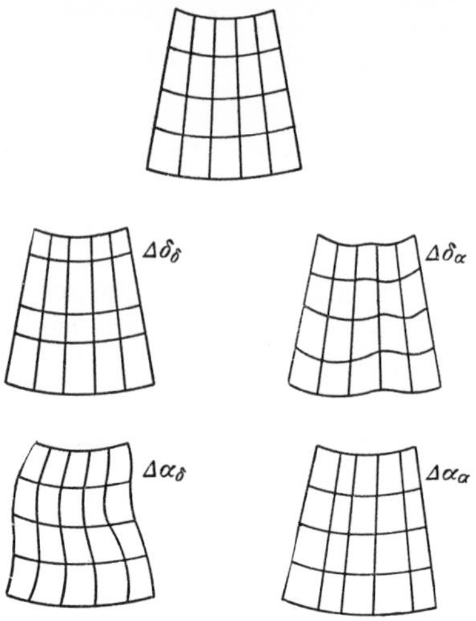

Fig. 65. Influence of systematic errors on the network of coordinates.

that $\Delta\alpha_\alpha$ and $\Delta\delta_\alpha$ are fairly well represented by a sinusoidal curve with a period of one year.

The systematic error of the type $\Delta\alpha_m$ depends upon the stellar magnitude and is a consequence of personal systematic errors. It can be well represented by the formula:

$$\Delta\alpha_m = a_0 + a(m - m_0) + b(m - m_0)^2,$$

where a_0, a, b are constant coefficients (b is usually negligibly small), m is the visual stellar magnitude and m_0 is some constant quantity. Declinations derived from visual observations, as a rule, do not have the error $\Delta\delta_m$. The photographic catalogs, however, do contain errors of the types $\Delta\alpha_m$ and $\Delta\delta_m$. Also characteristic of these catalogs are errors $\Delta\alpha_{sp}$ and $\Delta\delta_{sp}$ that depend upon the spectral class.

The system of proper motions μ_α and μ_δ in right ascension and in declination also contains its own systematic errors which depend upon the right ascension as well as upon the declination:

$$\Delta\mu_\alpha = (\Delta\mu_\alpha)_\alpha + (\Delta\mu_\alpha)_\delta,$$
$$\Delta\mu_\delta = (\Delta\mu_\delta)_\alpha + (\Delta\mu_\delta)_\delta,$$

and, with time, they change the systematic errors $\Delta\alpha_\alpha$, $\Delta\alpha_\delta$, $\Delta\delta_\alpha$ and $\Delta\delta_\delta$. Systematic errors of proper motions, although small in themselves, accumulate. Thus, with time, even a well-established fundamental system, with small systematic errors at the mean observing epoch, gradually loses its precision in a systematic way.

The systematic errors of a catalog cannot be directly determined because the absolutely correct coordinates of the stars of the catalog

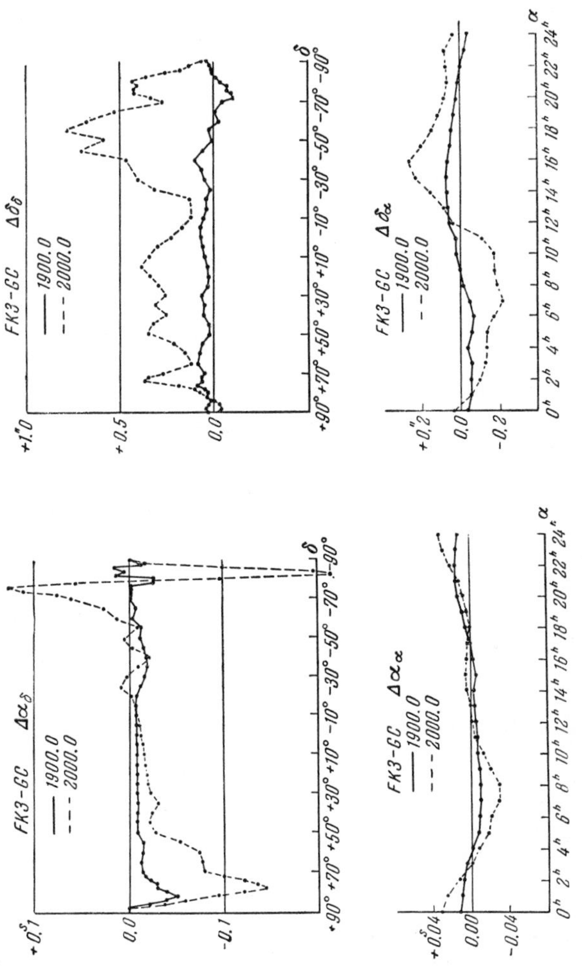

Fig. 66. Graphs of the systematic differences of the Catalogs FK3 and GC.

are not known; consequently the system of absolute stellar coordinates remains unknown. Moreover, indirect methods for deriving systematic errors based on comparisons between a specific catalog and many other catalogs are not free from certain restrictions. When we use separate catalogs, it is possible to examine the differences between the systematic errors of only two catalogs or, in other words, their *systematic differences*. Systematic differences of positions and proper motions are designated in the same way as the systematic errors. Systematic differences are obtained from comparison of catalogs.

In order to compare two catalogs, it is necessary to reduce both of them to the same epoch t_0. For this purpose corrections for the precession and proper motion are added to the stellar coordinates. The first correction is calculated for the time interval $(t_0 - t_i)$ where ti is the time of the catalog equinox, while the second is computed for the interval $(t_0 - t_i')$ where t_i' is the epoch of observation in the catalog. The coordinates of the common stars will now differ by the combined accidental and systematic errors of each catalog. Let us form the differences $\Delta\alpha = \alpha_1 - \alpha_2$ and $\Delta\delta = \delta_1 - \delta_2$ for stars which are included in each catalog.

The systematic difference ΔA of the zero points can be determined as the arithmetic mean $\overline{(\alpha_1 - \alpha_2) \cos \delta}$ of all the stars and is subject to elimination from the differences $\Delta\alpha$; otherwise it will enter into $\Delta\alpha_\delta$ as a constant shift. The average value of $\Delta\alpha$ for the equatorial zone is usually accepted for ΔA, which is then excluded from all $\Delta\alpha$.

By averaging $\Delta\alpha$ and $\Delta\delta$ according to declination zones, for example, in five-degree zones, we obtain the quantities $\overline{\alpha_1 - \alpha_2}$ and $\overline{\delta_1 - \delta_2}$ for the average declination of the zone. By plotting these quantities on a graph and drawing a smooth curve through the points, we shall obtain the values of $\Delta\alpha_\delta$ and $\Delta\delta_\delta$ for any declination. Sometimes a more formal analytical method for smoothing is used.

By eliminating the derived values $\Delta\alpha_\delta$ and $\Delta\delta_\delta$ from the coordinate differences of each star according to its declination, we obtain the differences $(\alpha_1 - \alpha_2) - \Delta\alpha_\delta$ and $(\delta_1 - \delta_2) - \Delta\delta_\delta$ which we average according to the right ascension hours; the quantities $\overline{\Delta\alpha - \Delta\alpha_\delta}$ and $\overline{\Delta\delta - \Delta\delta_\delta}$ are smoothed out graphically or analytically and are considered to be the systematic differences of the form $\Delta\alpha_\alpha$ and $\Delta\delta_\alpha$. Now, we may obtain the second approximation of $\Delta\alpha_\delta$ and $\Delta\delta_\delta$ by eliminating the differences $\Delta\alpha_\alpha$ and $\Delta\delta_\alpha$ from the initial differences $\Delta\alpha$ and $\Delta\delta$, and by averaging the corrected differences over again in the declination zones. In view of the small size of $\Delta\alpha_\alpha$ and $\Delta\delta_\alpha$, their second approximation is usually not determined.

The residual differences $\Delta\alpha - \Delta A - \Delta\alpha_\delta - \Delta\alpha_\alpha$ are distributed according to the stellar magnitudes for the determination of $\Delta\alpha_m$; the differences are usually combined within each half magnitude interval.

Systematic differences between the proper motions of two catalogs are obtained in a similar way.

Figure 66 represents examples of graphs of systematic differences $\Delta\alpha_\delta$, $\Delta\alpha_\alpha$, $\Delta\delta_\delta$, and $\Delta\delta_\alpha$, for the two fundamental catalogs FK3 and GC according to A. Kopff. The graphs show at a glance the order of values of the systematic differences for two first-class systems, as well as the effect of the systematic differences between the proper motions which make the system worse when going over to another more distant epoch.

78. Internal and External Accuracy of a Catalog with Respect to Accidental Errors

An estimate of the accuracy of an initial catalog which is based only on material contained in the catalog itself is called an estimate of its accuracy according to its internal agreement. A comparison of the positions of stars in a given catalog with the positions from other catalogs permits estimation of the external accuracy.

In the preparation of a catalog each star is observed, as a rule, at least twice. Very frequently, particularly in case of increase of small programs, each star may be observed up to five or ten times. This makes it possible to estimate accuracy of the catalog from the agreement between individual coordinate values obtained for one star. The systematic differences of the type "circle east minus circle west," "upper culmination minus lower culmination," etc., must be preliminarily excluded.

Before calculating the mean square error ε_α and ε_δ of the position of a star in the catalog, it is first of all necessary to determine the mean square error ε'_α and ε'_δ of one observation. For this purpose we form the differences $\Delta\alpha$ and $\Delta\delta$ between individual observed coordinates and their mean value for each star. The mean square error of one observation, particularly for right ascensions, is usually calculated for each declination zone in view of the dependence of this quantity on the declination. Strictly speaking, one should carry out calculations separately for groups of stars having the same number of observations; however, in practice we apply the formulas:

$$\varepsilon'_\alpha = \pm\sqrt{\frac{\sum \Delta\alpha^2}{N-m}} \text{ and } \varepsilon'_\delta = \pm\sqrt{\frac{\sum \Delta\delta^2}{N-m}},$$

where N is the common number of deviations $\Delta\alpha$ or $\Delta\delta$ and m is the number of stars.

The mean square error of the cataloged position is obtained from the relations $\varepsilon_\alpha = \frac{\varepsilon'_\alpha}{\sqrt{n}}$ and $\varepsilon_\delta = \frac{\varepsilon'_\delta}{\sqrt{n}}$ where n is the average number of observations of one star. There is a dependence of the quantities ε_α and ε_δ on the number of observations and on the declination.

The dependence on the number of observations is represented for both coordinates by the formula $\varepsilon^2(n) = \varepsilon^2(\infty) + \frac{\varepsilon'^2}{n}$ where $\varepsilon(\infty)$ is the mean square error for a large number of observations and can be determined from comparison with other catalogs. Such a dependence is not always very definite. The main reason for the existence of $\varepsilon(\infty)$ is the fact that $\Delta\alpha$ and $\Delta\delta$ actually contain non-eliminated systematic errors. Hence, the distribution of $\Delta\alpha$ and $\Delta\delta$ is not a normal distribution. The quantity $\varepsilon(\infty)$ is in some ways a measure of elimination of the systematic errors.

The dependence of the mean square error of one observation upon the declination appears very definitely in right ascension and can be adequately represented by the formula $\varepsilon_\alpha^2(\delta) = a^2 + b^2 \sec^2 \delta$, which shows the dependence of the accuracy of the obtained right ascensions upon the speed of guiding on the star. However, an exact proportionality between ε_α and the secant of the declination (i.e., the speed of the moving star) is not found. Apparently this is due to the image scintillation, which changes with zenith distance. Sometimes, in order to compensate for this effect, a term with the form $c^2 \sec^2 \delta \tan^2 z$ is introduced into the formula relating ε_α to δ. The accidental error in the determination of declinations depends on the zenith distance of the star, basically because of the refraction anomalies, the scintillation and the atmospheric dispersion, which make the image quality worse as one approaches the horizon.

Figure 67 shows, as an example, a graph of the variation of the mean square error of right ascensions as a function of declination.

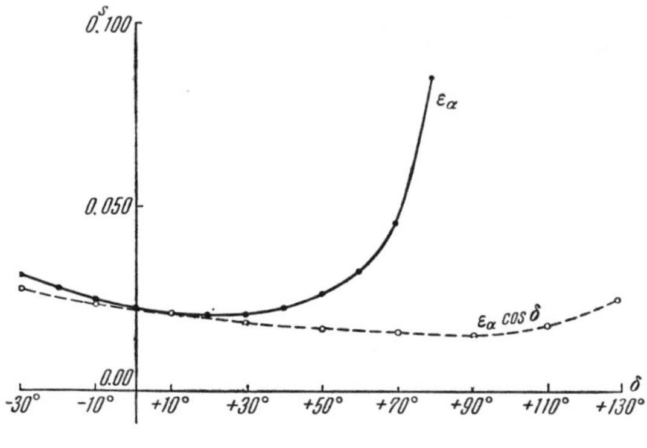

Fig. 67. Variation of the mean square error of right ascensions with the declination.

The mean square error of a single observation obtained with a meridian circle is roughly equal to $\varepsilon'_\alpha \cos\delta \approx \pm 0^s.02$ for right ascension, and $\varepsilon'_\delta \approx \pm 0''.4$ for declination in the first-class modern catalogs. Since the number of observations usually ranges from 4 to 8, the corresponding accuracy will then be $\varepsilon_\alpha \cos\delta \approx \pm 0^s.008$ to $\pm 0^s.010$ and $\varepsilon_\delta \approx \pm 0''.15$ to $\pm 0''.20$.

The accuracy of catalogs with respect to accidental errors is determined from external agreement by forming the differences $\Delta\alpha \cos\delta$ and $\Delta\delta$ of star coordinates obtained from two catalogs. It is essential that the two catalogs be reduced to the same system. The mean square difference $\varepsilon_{\mathrm{I-II}}$ for both coordinates is calculated from the usual formula $\varepsilon^2_{\mathrm{I-II}} = \frac{\sum \Delta^2}{n}$, where n is the number of common stars; moreover, the denominator is equal to n instead of $n-1$ since $\Delta\alpha$ and $\Delta\delta$ are displacements from the correct value of zero. On the other hand,

$$\varepsilon^2_{\mathrm{I-II}} = \varepsilon^2 + \varepsilon^2_{\mathrm{II}}, \tag{53}$$

where ε_{I} and $\varepsilon_{\mathrm{II}}$ are the mean square errors of the positions in the catalogs. If, for example, catalog I is thoroughly studied and its ε known with confidence, then it is possible to determine from the relation (53) that quantity $\varepsilon_{\mathrm{II}}$ which will characterize the external agreement of catalog II. As a rule, the mean square error from external agreement is larger than from internal agreement, which can be explained by the presence of the systematic errors which have not been eliminated from $\Delta\alpha$ and $\Delta\delta$.

79. Construction of a Fundamental System

The presence of accidental and systematic errors in catalogs renders the derivation of the most probable fundamental system fairly complex. As was shown above, the standard coordinate system is established in the form of a fundamental catalog containing the coordinates and proper motions for a number of stars obtained by combining all available absolute and relative catalogs.

The task of compiling the fundamental catalog can be separated into two parts: the derivation of a new, most probable system of coordinates and proper motions, and the improvement of the values of the coordinates and proper motions of stars which define the system, with respect to accidental errors. At the present stage of the problem, the fundamental system, as a rule, is not to be newly derived but only small corrections to an already existing system are to be determined. Therefore it is possible to begin by improving the fundamental catalog with respect to accidental errors, or to improve the system internally.

Let the fundamental catalog to be improved be K_0 and its equinox t_0. The star coordinates and proper motions, reduced to the new system, are designated provisionally by x_0 and μ_x. The catalogs to be used for improvement are K_1, $K_2, \ldots K_m$ with the equinoxes t_1, $t_2 \ldots t_m$ and observation epochs t'_1, $t'_2 \ldots t'_m$. These can be either absolute or relative catalogs. All of the coordinates in the catalogs K_i are reduced to the equinox and epoch t_0 by means of precession and proper motion corrections taken from the K_0 catalog.

By comparing each catalog K_i with K_0, systematic differences of various kinds are determined by the method described in Section 77. By eliminating these differences from the star coordinates of catalog K_i, this catalog can be brought to the fundamental system of the catalog K_0. The resulting coordinates x_1, $x_2 \ldots x_m$ of any star in the catalogs K_1, $K_2, \ldots K_m$ will differ from the coordinates x_0 in the catalog K_0 only by accidental errors. We designate the necessary corrections by Δx_0, Δx_1, $\Delta x_2 \ldots \Delta x_m$. In addition, there will be a correction due to the accidental error brought in through the proper motion, namely $\Delta\mu_x (t'_i - t_0)$, where $\Delta\mu_x$ is the correction for the accidental error of μ_x. Thus it is possible to set up for each star m equations of the form

$$x_0 + \Delta x_0 = x_i + \Delta x_i + \Delta\mu_x (t_0 - t'_i), \quad i = 1, 2, 3, \ldots, m. \tag{54}$$

Since the quantities Δx_i are random quantities, it is possible to consider them as the corresponding residuals to the individual conditional equations. Then, after solving the system of conditional equations

$$\Delta x_0 + \Delta\mu_x (t'_i - t_0) = (x_i - x_0), \quad i = 1, 2, 3, \ldots, m,$$

it is easy to obtain the value of Δx_i by substituting the determined unknowns Δx_0 and $\Delta\mu_x$ into (54).

The system of equations is to be solved with assigned weights, called weights with a random relation. These weights are assigned according to the accuracy of the catalogs with respect to accidental errors according to the formula

$$p_i = \frac{\varepsilon_0^2}{\varepsilon_i^2(n)},$$

where ε_0 is the mean square error of the unit weight and is assumed to be equal to $\pm 0''.45$ according to an estimate by Boss. The quantity $\varepsilon_i(\bar{n}) = \frac{\varepsilon_i'}{\sqrt{\bar{n}}}$ is taken as the first approximation of $\varepsilon_i(n)$, and $\bar{n}$ is the average number of observations of one star in the given catalog. The quantity ε_i' (the mean square error of a single observation according to the internal agreement is given for each catalog). After calculating the residuals Δx_i for stars with an equal number of observations k listed in the catalog K_i, it is possible to calculate $\varepsilon_i(k)$ by the formula

$$\varepsilon_i^2(k) = \pm \frac{\sum^{s} \Delta x_i^2}{s-1},$$

where s is the number of residuals Δx_i. Using the quantities $\varepsilon_i(k)$ calculated in this way, it is possible to extrapolate the quantity $\varepsilon_i(\infty)$. The final weights p_i are assigned according to the number of observations of a given star in the catalog by using the formula

$$p_i = \frac{\varepsilon_0^2}{\varepsilon_i^2(\infty) + \frac{\varepsilon_i'^2}{n}}.$$

Together with these are also found the final corrections Δx_0 and $\Delta \mu_x$.

In order to derive the corrections for the fundamental system of positions and of proper motions of the catalog K_0 only the absolute catalogs K_1, $K_2 \ldots K_k$ are used. The systematic differences $(K_i - K_0)$ have been obtained earlier as the result of comparing each of the catalogs K_i with K_0. These systematic differences are corrected for all the obtained systematic errors of the catalog K_i. The determination of the systematic corrections to the K_0 catalog is made separately according to the various forms of the systematic differences.

The resulting systematic difference between the catalogs K_i and K_0 (e.g., the difference $\Delta\alpha_\delta$) is considered as due to the systematic errors in positions and proper motions of the catalog K_0. Thus it is possible to represent $(\Delta\alpha_\delta)_i$ for the K_i catalogue in some declination zone in the form

$$(\Delta\alpha_\delta)_i = (\Delta\alpha_\delta)_0 + (\Delta\mu_\alpha)_\delta (t_i' - t_0),$$

where $i = 1, 2, 3, \ldots k$. By solving the system of k conditional equations which are weighted in a systematic way, the corrections $(\Delta\alpha_\delta)_0$ and $(\Delta\mu_\alpha)_\delta$ of the system of the catalog K_0 are determined. The derived systematic corrections are intercompared for all zones and smoothed out.

The weights are assigned in a systematic way after a detailed study of the catalogs. Attention is given to the methods of eliminating systematic errors: reversing the telescope during the observations, taking into account instrumental errors, and independent determination of the coordinates. Although the systematic weights are not calculated, there is a series of criteria allowing one to verify the objectivity of the assigned weights. Thus, for example, the quantity $\varepsilon_i(\infty)$ can serve as a measure of the degree to which the systematic errors of the catalog have been eliminated. Consequently, it is a quantity which is approximately inversely proportional to the weight in the systematic relation and has a random relation in the determination of the weights.

The remaining systematic errors are obtained in a similar way. The corrections of the type $\Delta\alpha_m$ are easily found and are eliminated from catalogs possessing these errors. The determination of the equinox correction ΔA is a problem to be solved separately.

The presented method for the derivation of the system of the fundamental catalog is a simplified scheme. Practical application of this method is made complex because the individual catalogs obtained by classical methods sometimes have large systematic errors and the derivation of the mean weighted system without determination and elimination of these errors does not give satisfactory results. The described method is essentially empirical and is not based on any long-established principles. It is used only for the purpose of compensating the systematic errors of the initial catalogs, and also under the assumption that the systematic errors are small and their discrepancy is of random character. However, this assumption is not true; systematic errors of individual catalogs may be of significant size (of the order of one to two seconds of arc).

80. The Problem of Improving the System of a Fundamental Catalog

At the present time, the improvement of the fundamental systems of coordinates is the most important problem in fundamental astrometry. The necessity for such improvement is confirmed by making comparisons between the best fundamental systems and new observations. Several ways for solving the problem are indicated. On the one hand, one tries to organize the observations and the reduction of the initial absolute catalogs in order that their systematic errors be at a minimum; on the other hand, new methods are developed for incorporation of catalogs into a fundamental system. Of primary importance are the methods proposed for improvement of the fundamental system of positions by observing minor planets and for improvement of the system of proper motions by observing extragalactic nebulae. Both of these methods, as well as new instruments, which differ from the classical instruments both in their structure and in principle, will be discussed in the following paragraphs.

The greatest inconvenience is caused by systematic errors in declinations, especially in the form of $\Delta\delta_\delta$. Almost all catalogs of declinations show a good agreement near the pole, but give large discrepancies near the equator. These differences are explained by the effect of the instrumental flexure, which is not completely taken into account, and by the incomplete elimination of the refraction; they sometimes reach one to two seconds of arc. Therefore, it is not surprising that the majority of suggestions concern the improvement of the declination system. By way of example, let us describe some of these suggestions.

a) By simultaneous observations with the meridian circle and the zenith-telescope of Talcott pairs of stars we obtain: with the meridian circle, the sum of the zenith distances of these stars from the difference of the circle readings; and with the zenith-telescope, the difference between the zenith distances of the same stars. These quantities are related to the declinations through the following expressions:

$$\left.\begin{aligned}(z_n+z_s)+(\rho_n+\rho_s)+(\Delta_n+\Delta_s)&=\delta_n-\delta_s,\\(z_n-z_s)+(\rho_n-\rho_s)+2\varphi&=\delta_n+\delta_s,\end{aligned}\right\}\qquad(55)$$

where ρ is the refraction. The quantity $(\Delta_n+\Delta_s)$, which is the effect of flexure in the meridian circle, contains only terms of horizontal flexure, since (see Section 26) it is equal to

$$a(\cos z_s-\cos z_n)+b(\sin z_s+\sin z_n)\approx 2b\sin z.$$

By determining b with the horizontal collimators, the flexure can be taken into account.

If one observes one and the same northern star also at its lower culmination together with a southern star, then we shall have two more relations:

$$\left.\begin{aligned}(z'_n+z'_s)+(\rho'_n+\rho'_s)+(\Delta'_n+\Delta'_s)&=180^\circ-\delta_n-\delta'_s,\\(z'_n-z'_s)+(\rho'_n-\rho'_s)+2\varphi&=180^\circ-\delta_n+\delta'_s.\end{aligned}\right\}\qquad(56)$$

From equations (55) and (56), it is possible to determine the three unknown declinations δ_s, δ'_s and δ_n as well as the latitude of the observatory. Consequently, it is not necessary to determine the zenith point from readings of the mercury surface and the latitude from observations of circumpolar stars. The amount of work in putting together this program and the necessity of taking into account the flexure limit the application of this method, which can be considered only as a check.

b) A comparison of declinations of stars obtained from observations both at northern and southern observatories was the method used by L. Boss in the derivation of a fundamental system. He represented the difference in the declinations in the form

$$\delta_S-\delta_N=(\rho_N-\rho'_N)K_N-(\rho_S-\rho'_S)K_S$$

where ρ'_N and ρ'_S are the mean refractions at the pole for each observatory, ρ_N and ρ_S are the accepted mean refractions for the star in the northern and southern catalogs, and K_N and K_S are unknown coefficients by which the accepted refractions must be multiplied in order to obtain the corrections to the declinations δ_N and δ_S. This appears to be a formal method since all the differences are attributed to refraction. However, as results show, this method improved the system of some catalogs. A better method, wherein one northern catalog is compared with a group of southern catalogs and, conversely, one southern catalog is compared with a group of northern catalogs, was used by Kopff in compiling FK3.

c) The method for forming the fundamental system of declinations from observations made at various latitudes was proposed by V. G. Shaposhnikov and is based on the principle of zenith symmetry: the meridian arcs which are symmetrically distributed with respect to the zenith and are equal in a given declination system, are equal in reality.

The declinations of the zenith point and of two points placed symmetrically with respect to it and having a zenith distance z for the observatory at the latitude φ_i will be φ_i, $\varphi_i - z$, $\varphi_i + z$. Let us designate the latitude correction by $\Delta\varphi_i$ and the systematic corrections of the zenith distances by Δz_n and Δz_s. Then, the systematic corrections $\Delta\delta_\delta^{(i)}$ for the declinations will be equal to:

$$\Delta\delta^{(i)}_{\varphi_i - z} = \Delta\varphi_i - \Delta z_s,$$

$$-\Delta\delta^{(i)}_{\varphi_i} = -\Delta\varphi_i,$$

$$\Delta\delta^{(i)}_{\varphi_i + z} = \Delta\varphi_i + \Delta z_n.$$

Multiplying the middle equation by two and adding it to the other two, we obtain the mathematical expression for the condition of the zenith symmetry

$$\Delta\delta^{(i)}_{\varphi_i - z} - 2\,\Delta\delta^{(i)}_{\varphi_i} + \Delta\delta^{(i)}_{\varphi_i + z} = 0,$$

since Δz_s is equal to Δz_n by the same principle.

The idea of this method can be clarified by an example of two observatories i and k located at the latitudes +30° and -30°. Since the systematic errors for circumpolar stars observed at two culminations are nearly equal to zero, it is possible to assume that $\Delta\delta^{(i)}_{+90°} = \Delta\delta^{(k)}_{-90°} = 0$. Therefore, the equations of the zenith symmetry for a zenith distance of 60° will be written in the form

$$\Delta\delta^{(i)}_{-30°} - 2\,\Delta\delta^{(i)}_{+30°} = 0,$$

$$\Delta\delta^{(k)}_{+30°} - 2\,\Delta\delta^{(k)}_{-30°} = 0.$$

By means of a comparison of catalogs i and k, the systematic difference for the declinations +30° and -30° will be known, which we designate by $\Delta\delta_\delta^{(i)-(k)}$:

$$\Delta\delta^{(k)}_{+30°} - \Delta\delta^{(i)}_{+30°} = \Delta\delta^{(i)-(k)}_{+30°},$$

$$\Delta\delta^{(k)}_{-30°} - \Delta\delta^{(i)}_{-30°} = \Delta\delta^{(i)-(k)}_{-30°}.$$

It is easy to derive from these four equations the systematic corrections for both catalogs corresponding to the declinations +30° and -30°. The conditions of the zenith symmetry should be combined for various zenith distances of stars observed at several observatories with reasonably chosen latitudes. Systematic corrections for catalogs can be obtained from this for a whole series of points along the meridian, and through this we may study the declination system of each catalog.

81. Improvement of the System of Positions with Observations of Minor Planets

Regular motions of members of the solar system with respect to stars are used for the determination of the systematic errors of catalogs. The use of solar observations was discussed in Section 65. For the same purpose observations of major planets (Mercury, Venus and Mars) are used. However, minor planets have many obvious advantages, one of them being that their observations do not differ in any way from star observations due to their small apparent diameter. The idea of observing minor planets was first suggested by the scientists of the Institute of Theoretical Astronomy of USSR and approved by the All-Union Astrometric Conference in 1932. Subsequent theoretical and practical work was undertaken in this direction in other countries as well.

It is more convenient to observe minor planets by a photographic method with an astrograph, rather than visually. This is due primarily to the fact that the minor planets are not very bright; besides this, the accuracy of the positions of minor planets obtained by the photographic method is three to five times higher than the accuracy of meridian observations. However, minor planets are observable only near opposition and have a comparatively long period of revolution. This can be compensated for by continuous observations of several minor planets, which will give us in a ten-year period a sufficient amount of data in order to derive refined effects of the systematic errors of the catalog.

From a general point of view, the idea of using minor planets is extraordinarily simple and does not differ from the well-known method of using major planets proposed by Newcomb. The coordinates of a minor planet α_N and δ_N, obtained by the photographic method, within the system of the catalog to be improved, are compared with previously calculated ephemeris coordinates α_e and δ_e. The differences $\alpha_N - \alpha_e$ and $\delta_N - \delta_e$ are due to the systematic errors of the catalog and the inaccuracies of the ephemeris of a minor planet. In order to decrease the effect of the latter, it is necessary to work out a good theory of motion for each of the observed minor planets. The errors in the ephemeris are due to the errors $\Delta\Omega$, Δi, Δn, Δe, $\Delta\omega$, ΔM_0 in the orbital elements of a minor planet, and the errors $\Delta\varepsilon$, $\Delta n'$, $\Delta e'$, $\Delta\omega'$, $\Delta M_0'$, in elements of the earth's orbit. Systematic errors of the catalog in the equatorial region may be represented in the form:

$$\Delta\alpha = \Delta A + a \sin\alpha + b\cos\alpha,$$

$$\Delta\delta = \Delta D + a' \sin\alpha + b' \cos\alpha,$$

where ΔA is the equinox correction, ΔD is the equator correction and the coefficients a, b, a', b' characterize the systematic errors of the types $\Delta\alpha_\alpha$ and $\Delta\delta_\alpha$.

Thus the systematic differences $\alpha_N - \alpha_e$ and $\delta_N - \delta_e$ can be represented in the form:

$$\left.\begin{aligned}
&\alpha_N - \alpha_e = \Delta A + a\sin\alpha + b\cos\alpha + \\
&+ f_\alpha(\Delta\Omega, \Delta i, \Delta n, \Delta e, \Delta\omega, \Delta M_0, \Delta\varepsilon, \Delta n', \Delta e', \Delta\omega', \Delta M_0'), \\
&\delta_N - \delta_e = \Delta D + a'\sin\alpha + b'\cos\alpha + \\
&+ f_\delta(\Delta\Omega, \Delta i, \Delta n, \Delta e, \Delta\omega, \Delta M_0, \Delta\varepsilon, \Delta n', \Delta e', \Delta\omega', \Delta M_0'),
\end{aligned}\right\} \quad (57)$$

where f_α and f_δ are functions known from the theory of planetary motion and relate the unknown element corrections to the orbits of the minor planet and of the earth.

Theoretical investigations have demonstrated that the unknowns ΔA and ΔD are determined with greatest certainty. However, the coefficients a, b, a', b' cannot be completely separated from the effect of rotation of the fundamental system of coordinates caused by the systematic errors of proper motions.

Observation of several minor planets is very expedient since it enables the observer to fill out evenly the whole equatorial belt with a series of observations. Among the fairly bright minor planets, the following are considered convenient for observations: (1) Ceres, (2) Pallas, (3) Juno, (4) Vesta, (6) Hebe, (7) Iris, (12) Victoria, (18) Melpomene, (39) Laetitia, (51) Nemausa and several others which were intensely observed in the past and whose motion is well known.

Minor planets should consequently be observed if possible along a large section of the orbit; thus only those minor planets are suitable for observation whose magnitude at opposition is not less than $11^m.0$. They can be observed from two to three months before and after opposition. Since a minor planet must be observed along its complete orbit, the observations must be made at several oppositions; this is needed in order to correct the elements of its orbit. The observations should cover at least two periods of revolution around the sun, that is, not less than ten years. Minor planets must be observed if possible with wide-angle astrographs. A large number of stars with known coordinates photographed on the same plate facilitates the tying in of positions of minor planets with the system of the catalog to be improved.

The solution of the set of equations (57), made up with observations of all minor planets, contains a large number of unknowns and should preferably be done with electronic computers. First, the corrections for the elements of the earth's orbit must be determined using all the equations since these corrections are common to all the equations; next, the corrections for the orbital elements of each minor planet are determined separately; and finally, after eliminating all these unknowns, the principal unknowns (that is, the corrections for the catalog system) are determined. One should point out that at present the theoretical problems of solving these equations and of determining all the unknowns is only at a preliminary stage of being worked out.

In conclusion, let us point out some of the features of photographic observations of minor planets.

The basic difficulty in photographic observations of minor planets is the non-existence of a suitable catalog of reference stars based on some first-rate fundamental system. The existing fundamental systems of bright stars are not applicable due to an insufficient number of stars and due to their great brightness. The coordinates of reference stars are therefore chosen from some intermediate system, for example, the system of the Yale catalogs. The reduction of observations of minor planets must be so organized that it will subsequently be fairly simple to take into account the systematic corrections for coordinates of reference stars, and thus to bring the minor planet coordinates into the system of the catalog which is to be improved. This method of reduction was proposed by Schlesinger and is called the method of dependences. It will enable one to estimate directly the influence of errors of the coordinates of each individual reference star on the coordinates of a minor planet and thus avoid a repeated reduction of the measurements.

In this method the following relationships between the ideal coordinates ξ and η of the object and the ideal coordinates ξ_i and η_i of the reference stars, obtained by an algebraic transformation from the Turner equations, are used:

$$\xi = D_1\xi_1 + D_2\xi_2 + \ldots + D_n\xi_n,$$
$$\eta = D_1\eta_1 + D_2\eta_2 + \ldots + D_n\eta_n,$$

where

$$D_j = \frac{x_j\left(x\sum y_i^2\right) - y_j\left(x\sum x_i y_i\right) + y_j\left(y\sum x_i^2\right) - x_j\left(y\sum x_i y_i\right)}{\sum x_i^2 \sum y_i^2 - \left(\sum x_i y_i\right)^2} + \frac{1}{n}$$

are numerical coefficients and are called "the dependences." They appear as functions of the measured coordinates x, y and x_i, y_i, and consequently can be calculated. The systematic correctionx $\Delta\alpha_i$ and $\Delta\delta_i$ of the coordinates of the reference stars are small. Therefore the corrections $\Delta\alpha$ and $\Delta\delta$ of the positions of a minor planet, when the coordinates of the reference stars are reduced to a new system, can be calculated from the relations

$$\Delta\alpha = D_1\,\Delta\alpha_1 + D_2\,\Delta\alpha_2 + \ldots + D_n\,\Delta\alpha_n,$$
$$\Delta\delta = D_1\,\Delta\delta_1 + D_2\,\Delta\delta_2 + \ldots + D_n\,\Delta\delta_n.$$

By correcting the coordinates of a minor planet by the quantities $\Delta\alpha$ and $\Delta\delta$, it is possible to obtain its position within the system of the new improved catalog.

The second difficulty to which one should pay attention during observation of minor planets is the effect of atmospheric dispersion on the photographic position of a minor planet. This error is caused by the difference between the spectral class of minor planets and the mean spectral class of the reference stars. Supplementary unknowns γ_0 and γ_1 are therefore included in equations (57) assuming that the dependence of the refraction coefficient on the spectral class is linear.

These supplementary terms for the right ascension and declination will respectively be equal to:

$$f'_\alpha\gamma_0 + f'_\alpha\Delta_{sp}\gamma_1 \text{ and } f'_\delta\gamma_0 + f'_\delta\Delta_{sp}\gamma_1,$$

where γ_0 and γ_1 are the unknowns. Now γ_0 is different for each planet since it depends on the deviation of the spectral class of the minor planet from the mean spectral class of all stars in the reference catalog, but γ_1 has a common value for all minor planets. The quantity Δ_{sp} is the deviation of the mean spectral class of reference stars on a given plate from the mean spectral class of all stars in the reference catalog. The functions f'_α and f_δ of the approximate coordinates of the minor planet are known.

The effect of the atmospheric dispersion can be greatly reduced if one chooses the reference stars within the spectral class limits F5 to K5, which in the average is close to the spectral class of minor planets. We should note that the KS7 catalog of faint stars contains stars which are rather homogeneously distributed with respect to spectral classes.

82. Improvement of the System of Proper Motions with Observations of Extragalactic Nebulae

The galactic rotation and the participation of the sun and stars in it disturb the inertial property of the coordinate system defined by the positions and proper motions of stars of the fundamental catalog. Thus it is impossible to construct a coordinate system free from the effects of space rotation from observations of stars which belong to our galaxy. The influence of the galactic rotation can be avoided if one refers the coordinate system to objects which do not participate in rotation. It was proposed in the nineteen thirties to observe extragalactic nebulae. The first observations of this kind were made by Soviet astronomers and then by US scientists.

These nebulae are distributed outside of our galaxy and, consequently, according to available data, have proper motions smaller than $0''.0001$ per year; this is significantly smaller than the accuracy of present day determination of proper motions of stars. Therefore, the directions toward the extragalactic nebulae can be assumed to be very constant within a period of a century. Their uneven distribution on the celestial sphere does create some complications. In the vicinity of the galactic plane, these nebulae are not visible due to the presence of dark clouds which absorb their light.

The problem of determining the systematic correction for proper motions of the stars of the fundamental catalog obtained from meridian observations in a given region of the sky is reduced to the derivation (from photographic observations) of the mean displacement of these stars relative to the extragalactic nebulae located in this region. If $\overline{\Delta\mu_*}$ is the average proper motion of a group of stars in the fundamental catalog and obtained from meridian observations, and $\overline{\Delta\mu_N}$ is the average displacement of these stars relative to the nebulae obtained from photographic observations, then the systematic correction within this region of the sky for the proper motions will be $\Delta\mu = \overline{\Delta\mu_N} - \overline{\Delta\mu_*}$.

In practice, this is accomplished by photographing extragalactic nebulae at two epochs and using the method described in Section 75. The proper motion of one or a group of nebulae is determined relative to the reference stars. The proper motions of the reference stars must be known in the system of the fundamental catalog to be improved. Since the measuring accuracy of the diffuse images of extragalactic nebulae is usually lower by a factor of two as compared with the accuracy of star measures, selection of several extragalactic nebulae within a given region is very useful.

Extragalactic nebulae suitable for astrometric purposes must have symmetrical images in order that the position of their centers will not depend upon the exposure time and the plate sensitivity. The choice of such extragalactic nebulae is simple when fainter extragalactic nebulae are observed. The limit of nebular brightness is set by the power of the present-day astrographs. With long-focus astrographs (for example, with a standard astrograph) extragalactic nebulae of $13^m.0$–$14^m.0$ are observable, whereas more powerful astrographs can photograph extragalactic nebulae down to $16^m.0$. It is very difficult to obtain proper motions free from the magnitude error which is the basic systematic error in the photographic determination of proper motions. Therefore, photographs must be taken with diffraction gratings in front of the objective.

The best program for observation of extragalactic nebulae was worked out at the Lick Observatory. These observers use a diffraction grating which gives images of the first-order spectral images which are four stellar magnitudes fainter than the central images of stars and take two exposures: one long exposure, in order to bring out nebulae, and a second short exposure, so that the reference stars of the eighth magnitude do not appear overexposed. By using such photographs it is possible to get intermediate stars of $12^m.0$ and use these to obtain proper motions of bright stars of $8^m.0$ relative to the faint extragalactic nebulae of $16^m.0$. This is done by measuring images having the same diameters and, consequently, basically free from the magnitude equation. Proper motions of $12^m.0$ stars relative to extragalactic nebulae are obtained from long exposures by measuring their first-order spectral images, which appear on the plate as $16^m.0$ stars. Now, in order to obtain the proper motion of $8^m.0$ stars of the reference catalog relative to extragalactic nebulae, it is necessary to determine the proper motion of $8^m.0$ stars relative to the intermediate $12^m.0$ stars and add it to the already-found proper motions of $12^m.0$ stars with respect to the extragalactic nebulae. The relative proper motion of stars between $12^m.0$ and $8^m.0$ are obtained from short exposures since the spectral grating images of bright stars are of $12^m.0$. We remember that the arithmetic mean of the measured coordinates of spectral images are taken to be the coordinates of the center of the bright central image.

In using long-focus astrographs with a small field, there is an insufficient number of catalog stars on the plate and therefore it is difficult to interconnect these stars with extragalactic nebulae. At the Pulkovo Observatory, A. N. Deutsch proposed a statistical method for interconnecting these measurements when the proper motion of extragalactic nebulae is determined relative to the background stars whose magnitudes are close to the magnitudes of extragalactic nebulae, or of the order of $14^m.0$. In order to connect the extragalactic nebulae with the surrounding catalog stars, the latter are photographed on separate plates, using diffraction gratings for reducing the brightness of catalog stars to the brightness of the background stars. From these plates, on which there are other background stars, the displacement of catalog stars relative to the background stars of $14^m.0$ are obtained. Assuming that the background stars in the vicinity of the nebulae on the average possess a constant motion (this is statistically fulfilled), the proper motions of catalog stars are determined relative to the extragalactic nebulae. This method does not seem to be rigorously astrometric; however, its application is justified since it is impossible to utilize directly the astronegatives obtained with long-focus astrographs for the derivation of proper motion of bright stars relative to extragalactic nebulae. This is possible by having recourse to supplementary photographs obtained with wide-angle astrographs and will determine from these the proper motion of reference background stars relative to catalog stars.

83. New Instruments and Methods of Fundamental Astrometry

The improvement of existing classical meridian instruments and the construction of essentially new instruments will contribute to improvement of the initial catalogs and eventually to an improvement of the fundamental system obtained through them. All the work in this direction has as its aim the elimination or decrease of systematic errors in observations and an increase in accuracy as far as accidental errors are concerned, as well as the simplification of the observations and their reduction. Great attention is paid to the automation of the process of observation and to making all the readings by means of photoelectric techniques and photography.

Among the foremost improvements are new methods of registering star transits. Thus the photoelectric method is successfully used in obtaining star catalogs at the Pulkovo Observatory. Several procedures have been recommended for the photographic recording of star transits by the use of a photographic plate placed in the focal plane of the meridian instrument. The moving star images leave their traces on the plate. If one constructs an apparatus activated by impulses from an astronomical clock, it is possible to produce a sequence of interruptions along the traces either by means of a displacement of the star image with a supplementary plane-parallel glass plate relative to the photographic plate, or by displacing the whole plateholder with the plate. Thus the photograph obtained on the plate is analogous to the record on a tape of a recording chronograph. By measuring the distances of the points of displacements from the focal network of the instrument threads impressed on the plate, and by knowing the exact time of each displacement, it is possible to obtain the transit time of the star across the central thread of the instrument. This method was worked out at the Göttingen Observatory; however, it is applied only to bright stars.

For observations of fainter stars a compensating motion equal to the star's velocity is imparted to the photographic plate, as a result of which the star image appears as a point. An interesting procedure was used at the Lund Observatory. The star image is held in one place through the rotation of a plane-parallel plate at the required velocity around an axis perpendicular to the drift of the star image. The glass plate is placed immediately in front of the photographic plate. In these methods for obtaining the moment of the star transit, it is necessary to know the exact times corresponding to several positions of the moving plateholder or the glass plate.

The second group of improvements includes the previously described methods of impersonal reading of the graduated circle and of the eyepiece micrometer. Attention is called to the suggestion to sum up the eyepiece micrometer readings in right ascension by means of an electronic counter and in declination by means of a device similar to the one used with an aircraft sextant.

Many instruments with new principles have been proposed for the determination of star coordinates. All these have some advantages and some disadvantages; however, as yet, only the so-called horizontal meridian instruments and instruments operating on the principle of observation at equal altitudes are used in practice.

Horizontal meridian instruments preserve the principle of classical meridian instruments. A round mirror is placed on the horizontal axis, instead of a tube, with its plane parallel to the horizontal axis (Fig. 68). The fixed southern and northern telescopes for observing the stars are placed horizontally in the meridian on special piers. In addition to observations, they are also used for determining the instrumental constants (collimation, departure from parallelism between the mirror plane and the axis of rotation, etc.). The mirror must have a diameter twice as large as the diameters of the objectives of the telescopes so that the apertures of the objectives will be fully used at different positions of the mirror. The mercury surface at the nadir and the mires is used with this instrument for the same purpose as in classical instruments.

Fig. 68. Working model of the horizontal meridian circle of the Pulkovo Observatory.

The use of a mirror eliminates errors connected with tube flexure; the fixed positions of the telescopes makes it possible to record star observations conveniently and to control the position of the instrument. Among the disadvantages of the horizontal meridian circle is the doubling of errors connected with circle readings due to the introduction of a mirror into the optics of the instrument, as well as the inaccessibility for observation of stars situated low above the horizon. The mirror surface must be of high quality because the used portions of the mirror differ for stars with different zenith distances.

The basic equation of the horizontal meridian circle for reducing the transit time of the star to the meridian can be written down in the following way:

$$\Delta T = i \cos(\varphi - \delta) \sec\delta + k \sin(\varphi - \delta) \sec\delta \pm (k - k') \sec\delta +$$
$$+ 2\gamma \cos\left(45^\circ \mp \frac{\varphi - \delta}{2}\right) \sec\delta,$$

where i and k are the inclination and the azimuth of the horizontal axis, k' is the azimuth of the line of sight of the horizontal telescope and γ is the angle of inclination of the mirror to the horizontal axis. The upper sign of the $\pm$ signs refers to observations with the south telescope and the lower, to observations with the north telescope. Corrections for the irregularities of the pivots Δi_{ζ} and Δk_{ζ} are taken into account with the coefficients of i and k. The reduction of declinations does not differ in any way from the reductions of observations made with the classicial meridian circle except for the necessity of doubling the readings and their corresponding corrections.

Operating horizontal meridian circles at present have been installed in Pulkovo and Ottawa. The Pulkovo horizontal meridian circle, finally complete after many years of work by L. A. Sukharev, should be especially mentioned. This instrument completely utilizes all the possibilities for achieving automation of the observing procedure. A photoelectric device, with an electronic counter for summing up the contact times, is used for recording the transit times; also, automatic photoelectric microscopes provide the average circle reading for all the microscopes.

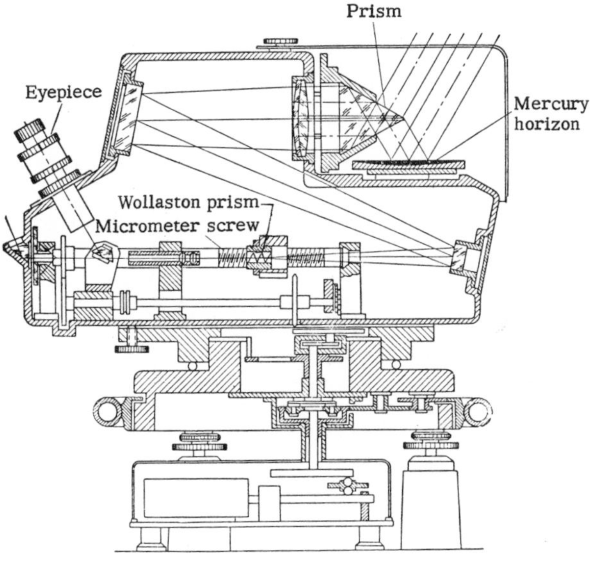

Fig. 69. Astrolabe of Danjon.

The prismatic astrolabe of Danjon is an instrument operating on the principle of equal altitudes (see Fig. 69). The star light falls directly on the upper face of an equilateral prism placed in front of the objective of the horizontal instrument tube and forms a star image in the focal plane of the objective (for the purpose of compactness, the optical axis of the instrument is broken by means of two mirrors). A second image is formed from the star light passing through the lower face of the prism after being reflected from the mercury horizon. At the instance of crossing the almucantar, close to $z_0 = 30°$, the two images coincide. By shifting the special prism with a micrometer screw having a contact drum, it is possible to keep the two images at a constant distance from each other and deduce from the recorded times the time T of crossing the almucantar. For observations at different azimuths, the instrument is rotated around its vertical axis. The basic formula of the prismatic astrolabe has the form:

$$\cos z_0 = \sin \varphi \sin \delta + \cos \varphi \cos \delta \cos (T + u - \alpha).$$

Thus this equation is convenient for the determination of the latitude and of the clock correction, as well as for the determination of star coordinates as long as φ and u are determined from reference stars, or obtained through smoothing out, or any other way.

E. Buchar of Czechoslovakia is constructing an apparatus (operating on the same principle and called a circumzenithal) in which the prism is replaced by two intersecting half-silvered mirrors, thus making it possible to alter the almucantar of the instrument.

The great inconvenience of the prismatic astrolabe or any instrument of this kind is the complicated reduction of observations and the impossibility of obtaining completely absolute determinations. The use of these instruments for determination of coordinates is yet to be justified by analysis of observational data.

CHAPTER XII

Catalogs of Positions and Proper Motions of Stars

84. A Short Historical Survey of Catalogs Obtained before the First Fundamental Catalog of Bessel

In concluding this book, in which we have described the instruments and methods of fundamental astrometry, we shall give a short survey of achievements in determining the positions and proper motions of stars from the birth of astronomy up to our time. The first problem confronted by early astrometrists was that of obtaining individual catalogs from observations, but afterwards, with the increasing demand for greater precision of the fundamental system, they turned to research of greater complexity which was based on observations made at many observations and with many instruments.

The first star catalogs were compiled by the Chinese. It is known that the catalog of Shih Shen in the fourth century B.C. contained 800 stars.

The ancient Greeks did considerable work and successfully determined star coordinates in terms of ecliptical latitude and longitude. Among the catalogs obtained by them, we may mention the catalog of 25 bright stars of Eudoxus (368-352 B.C.) and the catalog of 1022 stars by Hipparchus (123 B.C.). The compilation of the latter catalog led to the discovery of the existence of the precession of the equinoxes. The best known catalog from the ancient period is the Ptolemy catalog, which contained the coordinates of 1025 bright stars for the equinox of 138 A.D. Included in this catalog are Ptolemy's own observations as well as the observations of his predecessors. The accuracy of the catalogs for this epoch, which were made basically with armillary sphere observations, was $\pm 15'$. The Ptolemy catalog was applied in astronomical practice for several centuries, was reprinted many times and was reduced to other equinoxes (for 964—the catalog of Al-Suhi, for 1252—the Alfonsine Tables). As time went on, stars in the Ptolemy catalog were repeatedly reobserved in order to increase the coordinate accuracy, with the result that a whole series of catalogs appeared:

Ulugh Begh's catalog, containing 1017 stars for the epoch of 1437.5; Rothmann's catalog, containing 1004 stars for the epoch of 1594; and Tycho Brahe's catalog, containing 1005 stars for the epoch of 1601.

The catalogs of Hevelius (1564 stars for the epochs of 1661 and of 1701) were outstanding, possessing a threshold accuracy of $\pm 2'$ for observations with diopters. In these catalogs the right ascensions and declinations of stars were given for the first time. With the development of meridian principles of observation, the equatorial coordinates finally replaced other coordinate systems.

With the introduction of new principles of observation and new instruments, and especially with the replacement of diopters by the astronomical telescope, a radical increase in accuracy of the new catalogs was achieved. The Greenwich observatory, organized at the end of the 17th century, established itself as the principal center of positional astrometry. The Greenwich observatory retained this position up to the forties of the 19th century when the work of the Pulkovo observatory had developed. The Greenwich astronomers Flamstead, Maskelyne and particularly Bradley, increased the accuracy of coordinate determinations up to the present-day level. The accuracy of the Bradley catalogs for the epoch of the year 1755 is equal to $\pm 0\ .16$ in right ascension and to $\pm 1''.3$ in declination.

All the star catalogs before Bradley's have only historical interest in fundamental astrometry due to their low accuracy. Catalogs based on Bradley's observations have retained their significance to this day and have been used for the derivation of proper motions in constructing a number of fundamental systems.

With the establishment of the Pulkovo observatory in Russia, a leading observation center in fundamental astrometry was formed. The first-class catalogs of Pulkovo together with the Greenwich catalogs and the catalogs of other observatories formed the basis of modern fundamental systems. For the southern sky an important part was and still is played by the catalogs of the Cape observatory.

85. The First Fundamental Catalogs. The Newcomb Catalogs

The "Koenigsberg Tables" of Bessel should be considered as the first fundamental catalog compiled from independent observations. They contain the mean and the apparent coordinates of 36 Maskelyne equatorial stars and of the two circumpolar stars for the time interval between the years 1750.0 and 1850.0. The positions and proper motions of stars were derived by Bessel from his own observations (the catalogs for the years 1815 and 1825) and from observations by Bradley; for this purpose the latter were specially re-reduced by Bessel. The fundamental system derived by Bessel was used up to 1861 as a basis for the publication of data in the "Berliner Jahrbuch."

An improvement of the system of the "Koenigsberg Tables" was made by Wolfers, who used (in addition to the catalogs used by Bessel) twelve new catalogs: five in right ascension and seven in declination. In forming the new fundamental system only the mean differences between the catalogs were taken into account. As a result, a fundamental system was obtained from 47 stars and published under the name "Tables for reducing astronomic observations from 1860 to 1880'' and was used with the "Berliner Jahrbuch" up to the year 1883.

The fundamental system of coordinates accepted for the "Berliner Jahrbuch" was retained for a long time and advantageously distinguished this yearbook from the English and French yearbooks; the system of the first of these underwent a change with the appearance of each new Greenwich catalog, while the system of the second one was far from being perfect even for that not exacting period.

During the seventies of the last century Newcomb, Auwers and Boss commenced their extensive work, work which ultimately led to the formation of the modern fundamental systems and to the development of methods for investigation of systematic and accidental catalog errors.

The first catalog of Newcomb, designated by N_1, was published in 1872 and contained right ascensions of 32 equatorial clock stars (30 Maskelyne stars, Sirius and Procyon). This catalog was compiled in two stages. First, 12 best absolute catalogs were used in preparing a preliminary system. Then the final catalog was formed from investigation of 14 supplementary catalogs. The derivation of the equinox correction was very impressive, since it was obtained by an original method from solar observations for the period from 1750 to 1869. The equinox correction and the system $\Delta\alpha_\alpha$ of this catalog (N_1) was used in subsequent catalogs for over several decades because of their great accuracy.

The fundamental catalog N_2 of declinations and right ascensions of 1257 stars appeared in 1898. The right ascensions were corrected with the derived corrections $\Delta\alpha_\delta$. For the declination system the Boss system was accepted after taking into account the correction for the equator which Newcomb derived from observations of major planets using his own method.

86. The Undertaking of AGK and Its Fundamental Basis

In 1863, Argelander at the Bonn Observatory compiled an atlas and a photometric survey of 324,188 stars down to $9^m.5$ for the northern hemisphere between +90° and -2° in declination. The star numbers from the catalog with approximate positions for 1855.0 (appended to the atlas of the "Bonn Survey") are, some 100 years later, the standard designations of stars in the northern hemisphere. They consist of the letters BD (which are the initial letters of the name "Bonner Darchmusterung"), of the declination degree of the zone and of the star number within that zone; for example, for Polaris, its number from BD would be BD + 88°8, and for the star γ Vir, BD - 0°2601.

An International Astronomical Society founded in Germany in 1863 organized the first international cooperative undertaking to determine accurately the positions of all stars listed in the BD catalog down to $9^m.0$. This undertaking was named The Catalogues of the Astronomical Society: AGK (Katalog der Astronomischen Gesellschaft). Stars between the declination +80° and -2° were assigned to the participants by zones, which is the reason the AGK catalogs are sometimes referred to as the zone catalogs. Relative star coordinates were derived from at least two observations with meridian circles of twelve observatories in six countries. In Russia, the Kazan and Nikolayev observatories participated in this work.

For the purpose of homogeneity of these zone catalogs it was necessary to establish a fundamental catalog of reference stars. This catalog, published in 1879 and designated as FC, was called the first fundamental catalog of Auwers; the system was designated as A_1. In it are listed 539 stars, whose positions were derived from eight catalogs. Much data was taken from the observations of the Pulkovo Observatory, (namely, the catalogs Pu 1845, Pu 1865, and Pu 1871) as well as from Greenwich Observatory observations (published in the catalogs Gr 1861 and Gr 1872). These catalogs were reduced to the system of Pu 1865 after a study of the systematic differences of all kinds. Thus the system of A_1 is identical with the Pu 1865 system, which coincided in right ascensions with the system of the Newcomb catalog N_1 with respect to errors of the types ΔA and $\Delta\alpha_\alpha$, but is an independent system according to the errors of the $\Delta\alpha_\delta$ type, and in declinations. The proper motions were obtained by comparing one catalog, the Gr 1872, with the newly reduced Bradley catalog, and were therefore not very reliable. The FC system was used in the "Berliner Jahrbuch" from 1883 until 1906.

At the end of the eighties, the BD photometric survey was extended to the southern sky down to the declination -23°. At the Observatory of Cordova the CD survey was made, extending from -23° to -90°. At the Cape Observatory a photographic survey (CPD) was obtained, extending from -19° to -90° in declination. In connection with the extension of zonal observations to the southern sky, Auwers compiled the southern fundamental catalogs A_{303} and A_S. The catalogs A_{303} contained 303 stars with declinations between -2° and -23°, whereas the catalog A_S listed 499 stars between -20° and -82°.6. A significant role was played by the catalogs of the Cape and Cordova observatories in completing the above catalogs. The systems of right ascensions in the southern fundamental catalogs coincided with A_1 according to the errors of the type ΔA and $\Delta\alpha_\alpha$, whereas with respect to the errors of the type $\Delta\alpha_\delta$ and their declination systems these were independent systems.

The catalogs A_1, A_{303} and A_S were provisional fundamental catalogs, the incorporation and improvement of which results in the NFK catalog (Neue Fundamental Katalog). This catalog was published in 1907, listing the positions of 925 stars for the epochs 1870.0 and 1900.0. It was accepted for the "Berliner Jahrbuch" up to 1940 when it was replaced by the third fundamental catalog, FK3.

The undertaking of AGK was completed with the release, in 1910, of fifteen catalogs with positions of 144,000 stars for the 1875.0 epoch. An extension of observations to the southern sky down to -23° in declination added five catalogs for the 1900.0 epoch, which were completed around 1924. The accuracy of the AGK catalogs amounts to about $\pm0''.50$ in declination and correspondingly $\pm0.030 \sec \delta$ in right ascension. More recently, zonal catalogs for the remaining part of the southern sky were completed from observations at Cordova (from -23° to -52° and from -82° to -90°) and at La Plata (from -52° to -82°). As has been shown by the most recent observations, the AGK catalogs have large systematic errors.

87. The Fundamental Catalog of Boss. The System of Eichelberger

The work of Boss commenced with the compiling of the fundamental declination catalog which was published in 1879 and contained 500 stars; these were basically necessary for the determination of latitudes on the territory of the USA. The system of this catalog, which is usually denoted by B_1, was derived on the basis of 32 absolute catalogs observed between 1821 and 1872. An extension of this work led to the compilation of the B_S catalog in 1898 (179 southern stars between -20° and -50° in declination) whose right ascension system was based on the Newcomb system N_1 while the declination system was based on the system of the preceding catalog B_1. The combination of these catalogs led in 1903 to the creation of a fundamental catalog of 627 stars distributed over the whole sky. The system of the catalog B_{627} was independent with respect to $\Delta\alpha_\delta$ and $\Delta\delta_\delta$ errors, but in all other respects it repeated the N_1 and B_1 systems. The B_{627} catalog was the basis for the solution of the main problem proposed by Boss, namely to create a fundamental catalog with the greatest possible number of stars, including fainter stars. This openèd the possibility for the solution of such problems as the determination of the solar motion in space, the galactic rotation, the precession constant, all of which required knowledge of the positions and proper motions of a large number of stars.

To solve these problems, a list of all stars down to $7^m.0$ was compiled; this was later extended to include several thousand fainter stars. Thus it was necessary to extend the fundamental system B_{627} to a larger number of stars and to reobserve the stars, mostly rather faint, which at that time did not have a sufficient number of accurate positions. For the solution of the first part of the problem, a preliminary fundamental catalog (PGC) containing 6188 stars was constructed in 1910 on the basis of 82 catalogs, which reproduced the B_{627} system. The second part of the problem was solved by organizing meridian observations in the periods 1907-1908 and 1911-1918 with the meridian circle of the Albany Observatory (USA), and in the period 1909-1911 at San Luis (Argentina) where the same instrument was transferred for observations of southern stars. Thus two catalogs were obtained: the Albany catalog containing 20,811 stars; the San Luis catalog containing 15,333 stars.

After this, the compilation of the GC catalog (General Catalog) was started. By using essentially the PGC stars, a new and improved fundamental system was derived. This system was extended to all the remaining GC stars by reducing to the new system all the catalogs used. Then the positions and proper motions of all stars were improved with respect to accidental errors. The most difficult task was to improve the PGC system.

Let us consider this procedure in general outlines. No special investigations of the equinox correction ΔA were conducted. Since the system was based on the N_1 system, the correction to the Newcomb equinox was accepted; it was based on investigations of Kahrstedt and Morgan, who derived, respectively, the values: $\Delta A_{N_1} = 0^s.045$ and $\Delta A_{N_1} = 0^s.035$ for the 1900.0 epoch. An average correction $\Delta A_{N_1} = -0^s.040$ was applied to the PGC catalog. The systematic differences $\Delta\alpha_\alpha$ between each catalog and PGC were represented by a series

$$\Delta\alpha_\alpha = a_0 + a \sin\alpha + b \cos\alpha + c \sin 2\alpha + d \cos 2\alpha,$$

and the variation of each coefficient with time was studied after it was derived. The catalog was not corrected for this type of error since Boss assumed that the cause for this error was the uneven rotation of the earth. The $\Delta\alpha_\delta$ and $(\Delta\mu_\alpha)_\delta$ corrections for the PGC catalog were derived from 95 catalogs and taken into account. The corrections of the type $\Delta\delta_\alpha$ were investigated in separate zones and represented in the form:

$$\Delta\delta_\alpha = a \sin\alpha + b \cos\alpha,$$

but were not taken into account in forming the GC catalog due to large errors in the determination of coefficients. For the derivation of $\Delta\delta_\delta$, 93 catalogs were used after they were preliminarily corrected for instrumental and personal errors obtained from an analysis of the catalogs themselves. Corrections were introduced to the accepted refraction; they were determined from a comparison between northern and southern catalogs. After this, the corrections $\Delta\delta_\delta$ and $(\Delta\mu_\delta)_\delta$ were determined for the PGC catalog and these were also taken into account in compiling the GC.

When the new system had been derived, all catalogs were reduced to this system, and corrections of positions and proper motions of all GC stars for accidental errors were obtained.

The GC catalog which was published in 1937 contains positions and proper motions of 63,342 stars for the 1950.0 epoch and is based on 228 catalogs.

The accuracy of the GC catalog is uneven. Coordinates of two-thirds of the stars have a mean square error larger than the mean square error of the unit weight $\pm 0''.45$. As has been shown by new series of observations, the GC system has unaccounted errors $\Delta\alpha_\alpha$ and $\Delta\delta_\alpha$. The system of proper motions of stars has large errors caused by insufficient number of observations of faint stars; this leads to rapid deterioration of the system with time. Among the defects of GC one should also mention the uneven distribution of stars on the celestial sphere.

Almost all existing catalogs have been compared with the GC system. Systematic differences for 228 catalogs are given in the GC catalog and in the Gyllenberg tables for 108 more catalogs. This situation, together with the large number of stars in the GC catalog, enables one to bring within its system a majority of observations. Hence the GC system plays an important part, especially in stellar astronomy where it is adopted as the basic coordinate system for investigations of proper motions.

In concluding this section, let us give a short description of the Eichelberger fundamental system proposed during the twenties of this century when the accuracy of the existing NFK and PGC systems already did not satisfy the practical needs and the new systems FK3 and GC were only being set up. The catalog of positions and proper motions for 1504 stars for the 1925.0 equinox was put together by Eichelberger on the basis of four catalogs, two Washington and two Cape catalogs compiled in the 20th century. The proper motions were obtained from the PGC proper motions, for which were then derived the systematic corrections. The position of the Eichelberger catalog had small errors near the catalog epoch; however, due to poor proper motions, its accuracy quickly decreased with time. Nevertheless, this catalog was widely used in geodetic and astronomical practice up to the completion of the FK3 system.

88. The Fundamental Catalog FK3

Following the decision of the Astronomische Gesellschaft in 1924, a revision of the NFK system was undertaken in connection with the photographic reobservation of the AGK zone catalogs. A catalog of faint reference stars was necessary for this work; hence, there arose the problem of forming a new fundamental system and of obtaining a catalog of faint stars in this system. The new fundamental system was derived by Kopff in the early thirties and was designated as FK3. Initially an improvement was made of individual positions and proper motions of stars in the NFK system. Corrections for the coordinates and proper motions depending on accidental errors were derived for the northern hemisphere using some twenty catalogs; for the southern hemisphere, the catalogs of the Cape Observatory and catalogs of Cordova and San Luis were used. In the majority of cases the obtained corrections appeared to be small.

The improvement of the NFK system was made using 77 catalogs, the majority of which were published in the period between 1900 and 1930. The systematic corrections of the type $\Delta\alpha_\delta$ for the NFK system were derived using 20 catalogs; a good agreement was obtained for the systematic differences between all catalogs and the NFK system in the declination zone ±70°. For the circumpolar stars the $\Delta\alpha_\delta$ corrections were especially investigated by bringing in additional observations. The correction of the type $\Delta\alpha_\alpha$ was investigated in separate zones. The complex character of its change was established, thus showing a large increase in amplitude with an increase in the declination of the zone. The magnitude equation $\Delta\alpha_m = 0^s.0062\ (m - 4.0)$ was shown to be present and was calculated using 15 catalogs of the equatorial zone. The equinox correction for the NFK system was taken to be $\Delta A = 0^s.048$ according to the investigations of Kahrstedt. For the derivation of the systematic correction $\Delta\delta_\delta$, in addition to stellar catalogs, observations of the sun and of the major planets were included which were used for the derivation of the declination corrections of the individual equatorial catalogs. Further improvement of the declination system was achieved by combining each of the northern catalogs with a group of southern catalogs, and the other way around. The $\Delta\delta_\alpha$ correction was derived from 14 catalogs for the northern sky and from 6 catalogs for the southern sky. A great improvement in the method of forming the FK3 as compared with the GC system proved to be the empirical representation of the systematic differences by smoothing out the curves obtained from a direct comparison of the catalogs. In contrast with this method, in setting up the GC system, simplified analytical formulas were used to represent the changes of the systematic differences.

In order to improve the system of proper motions catalogs with epochs not earlier than 1845 to 1850 were utilized, but only those with extended series of observations using the same instruments (Pulkovo, Greenwich, Cape and some others). In order to improve the system of proper motions in declination solar observations and observations of minor planets were added.

The FK3 catalog contains 373 fundamental and 662 supplementary stars, distributed over the whole sky, for the 1950.0 equinox. The mean square errors of positions and proper motions of the FK3 catalog are small, particularly around the middle epoch of observation (about 1900.0). Thus, the right ascension errors in the equatorial zone for the epochs 1900.0 and 1950.0 are, respectively, equal to $\pm 0^s.002$ and $\pm 0^s.005$, and the declination errors are $\pm 0''.03$ and $0''.08$. Errors of the centennial proper motions are equal to $\pm 0^s.010$ and $\pm 0''.14$. The FK3 catalog is a first-class catalog, and, with respect to systematic errors, surpasses the GC catalog. Comparisons with new sets of observations are an obvious evidence of this. In accordance with the resolution of the Paris meeting of the International Astronomical Union (in 1935), all almanacs were transferred to this system; thus, the FK3 system became the basic fundamental system for astrometric and geodetic work for the whole world. The apparent places of the FK3 stars were published annually in London and, since 1960, have been published in Heidelberg.

89. The Extension of the FK3 System to Fainter Stars. The Undertaking of the AGK_2. The Fundamental System N_{30}

Toward the end of the twenties of this century, work on the photographic reobservation of the AGK zone catalogs was started. It was decided to obtain the catalog of the necessary reference stars by extending the FK3 system to fainter stars. For this purpose differential observations of faint stars with the meridian circles of seven observatories (Babelsberg, Bergedorf, Bonn, Wroclaw, Heidelberg, Leipzig and Pulkovo) were organized in the period 1928–1932. The final combined catalog, based on the FK3 system, contains 13,747 stars between $7^m.5$ and $9^m.0$, evenly distributed over the northern sky down to $-4°$ of declination. This catalog, designated as AGK_2A, reproduces well the FK3 system for fainter stars and has a high degree of accuracy: $\varepsilon_\alpha \cos\delta = \pm 0^s.009$, $\varepsilon_\delta = \pm 0''.18$.

Photographic reobservation of the zone catalogs was made at three observatories (Bergedorf, Bonn and Pulkovo) using similar types of astrographs. These instruments had a scale of $100''$ per millimeter, which corresponds to a focal length of 206 cm, and their objective diameters were 160 mm. A plate of 20×20 cm thus provided the image of a region equal to 25 square degrees on which there were, on the average, 15 AGK_2A reference stars. Consequently, with the reduction of the observations made between 1928 and 1932, the AGK_2 catalogs of Bergedorf and Bonn were issued containing 180,000 stars of the northern sky in the FK3 system of positions between $+90°$ and $-2°$ of declination. The mean square error of each coordinate in the AGK_2 catalogs is close to $\pm 0''.15$. The Pulkovo Observatory also published a star catalog for the polar zone from observations with the zone astrograph. Work on photographic reobservation of the zone catalogs has also been conducted at the Yale and Cape observatories and is still in progress there.

In concluding the description of existing fundamental systems let us say a few words about the fundamental catalog N_{30} of Morgan. This catalog was compiled in conjunction with the work of studying planetary motions and was formed by using only the observations of this century, or, more precisely, those catalogs which had not been used in compiling the FK3 and GC systems. More than 70 catalogs were used which had their middle epochs of observation between 1920.0 and 1950.0, and approximately 30 of these catalogs were entirely absolute. The proper motions of N_{30} were obtained from a comparison between the mean positions in N_{30} (1930.0) and in GC (1900.0) were improved with respect to systematic errors. The N_{30} catalog was issued in 1952; it contains 5268 stars more or less evenly distributed over the sky and includes all the stars from FK3 catalog as well as the stars of the Backlund-Hough list; 781 faint stars in the N_{30} are not listed in the GC catalog. The accuracy of the N_{30} catalog with respect to accidental errors is somewhat smaller than the GC accuracy; however, with respect to systematic errors, N_{30} is better than GC since more accurate modern series of observations were used. The latter situation helped to uncover the systematic errors not only in the GC system but also in the FK3 system, particularly in the southern hemisphere.

90. Photographic Catalogs

Mass determination of star coordinates by differential (visual and photographic) methods began at the end of the last century. The AGK and AGK_2 catalogs were discussed in the previous sections. We shall now describe other efforts in this direction.

The first international astrophotographic conference, meeting in 1887 in Paris, adopted the resolution to make a map of the whole stellar sky by photographic means. In order to accomplish this enormous work astrographs of the same type, the so-called standard astrographs, were built and the task was distributed among 18 observatories. With an objective diameter of 34 cm and a focal length equal to 3.4 m (scale $1'$ per mm), a $2° \times 2°$ area was obtainable on 16×16 cm plates. Since the image quality at the plate edges was not satisfactory, the sky was photographed in such a way that each photograph in a series half overlaps the adjacent photographs. On the basis of all these astronegatives it was planned that maps of the sky would be compiled for stars down to $14^m.0$ with the images obtained during long exposures. Accurate positions of all stars down to $11^m.0$ were to be obtained with short exposures. This voluminous work is still not finished, although certain steps are being taken toward its completion. The nonexistence of a general reference system of coordinates is a major handicap in this undertaking.

In 1909 an international astrophotographic conference in Paris made an attempt to organize the work of constructing a general reference system for the "Star Map" catalogs. The so-called Backlund-Hough star list was accepted for observation, which in its final form contained 3064 stars. It included the star list of the Pulkovo catalogs of 1900 and 1905, as suggested by Baklund, as well as the list of stars of the southern hemisphere compiled by Hough. In addition, a series of supplementary stars was included for the purpose of an even distribution of stars across the sky. On the basis of the catalogs which were to be obtained it was intended to compile an unusually extensive reference star catalog in view of the small field of the standard astrographs. This task has not been fully accomplished; however, stars of the Baklund-Hough list were observed at nine observatories: Algiers, Babelsberg, Washington, Greenwich, Nikolayev, Paris, Pulkovo, Uccle and Cape of Good Hope. The larger part of observations of stars to the north of $-10°$ in declination was incorporated in two combined Pulkovo catalogs. This combined catalog of declinations contains 1631 stars whose coordinates are obtained as the weighted average from nine separate catalogs after having reduced these to the system of the 1915.0 Pulkovo catalogs. The right ascension catalog, containing 1769 stars, was derived for the 1925.0 epoch as the average of 7 catalogs.

In 1914, Schlesinger at Yale University (New Haven, USA) commenced photographic reobservation of the zone catalogs with wide-angle astrographs, one of which had a field of 140 square degrees on a 48×58 cm plate and had a focal length of about 2 m. At present, positions for 150,000 stars in the $-30°$ to $+30°$, $+50°$ to $+60°$ and $+85°$ to $+90°$ zones have been obtained with a mean square error of $\pm 0''.15$. A defect in compiling the Yale catalogs is the insufficient accuracy in position of the reference stars and the non-homogeneity of their systems for different declination zones, since the reference stars were obtained with different meridian instruments. As a result, in individual cases systematic error in the Yale catalogs is as great as one second of arc. By a direct comparison of the coordinates in the Yale catalogs with those in the AGK catalogs, the proper motions of stars were obtained; these have a low accuracy because only two pairs of catalogs were used.

Photographic zone catalogs for the sky south of $-30°$ of declination are compiled from observations of the Cape Observatory; however, this work has not yet been completed.

Thus, as of the present time, there exists a large collection of accurate positions for 300,000 stars down to $9^m.0$. Sky maps contain very rich data for stars down to $11^m.0$. It is the immediate task of astrometrists to bring all these incomplete observations into one system, primarily for the purpose of setting up a homogeneous reference system of positions and proper motions convenient for any photographic determinations whatsoever.

91. Proper Motion Catalogs

Proper motions are derived from meridian observations of stars at different epochs, similarly to the derivation of stellar proper motions given in the fundamental catalog. Meridian observations are sometimes combined with photographic observations of stars reduced to the same system. These determinations give proper motions in the system of the meridian catalog and are sometimes referred to as absolute determinations. More precisely, one should understand the term "absolute" in the sense that it is attached to an inertial system.

The relative proper motions (relative to a group of reference stars) are obtained from photographic observations by comparing plates obtained at different epochs (see Section 75). The reduction of relative proper motions to absolute, that is, bringing them into the system of a given fundamental catalog, is possible only if a sufficient number of stars with known absolute proper motions are present on the plate, which happens in practice very rarely. Therefore the reduction of relative proper motions to absolute is done using corrections for the parallactic motions of reference stars and for the galactic rotation based on statistical data, e.g., the tables of Van Rhijn and Bok.

The problem of the precise reduction to absolute proper motions for any stars whatever will be solved when the absolute proper motions for 180,000 faint stars (of the northern sky) of the AGK_3 list are determined; the latter can then be used as reference stars for any photographic determination of proper motions. In order to solve this problem, it is necessary to establish a meridian catalog for the reference system AGK_3R, which is one of the basic problems in modern fundamental astrometry which the astrometrists must solve before the end of the 20th century.

The accuracy of the absolute proper motions obtained will be increased as knowledge of the systematic errors in the positions used becomes more precise, as random errors in positions decrease and as the interval between epochs is increased. It is usually assumed that the average accuracy of the yearly proper motions of the bright stars is $\pm 0''.001$ and for the faint stars, $\pm 0''.01$, which is determined basically by the number of observations and by the maximum time interval between the epochs. Fainter stars were little observed in the last century and hence their proper motions are as a rule less accurate. One should consequently point out that the increasing accuracy of observations in the 20th century leads to the fact that the accuracy of proper motions derived from two observations with the mean epochs 1800.0 and 1900.0 is found approximately to be the same as the accuracy of modern observations with mean epochs around 1900.0 and 1930.0. The accuracy of the determination of relative proper motions from photographic plates depends upon the focal length of the astrograph and upon the interval between the epochs of the two astronegatives; it is found to be on the average close to $\pm 0''.005$ to $\pm 0''.010$.

Among absolute proper motions one should mention, in the first place, the proper motions of stars of the fundamental catalogs FK3 (1535 stars), GC (33,342 stars) and N_{30} (5268 stars), which were obtained from meridian observations of these catalogs, the proper motions of faint stars of the GC catalog are least accurate due to an insufficient number of observations. The mean square error of the proper motions for bright stars in all these catalogs is of the order of $\pm 0''.001$ and for faint stars of the order to $\pm 0''.01$.

Absolute proper motions for a large number of stars down to $10^m.0$ are obtained by comparing new photographic positions with old visual ones. Such work has been done at the Yale University, where proper motions of 144,000 stars listed in the AGK catalogs were obtained from the differences between the photographic positions of the Yale catalogs and the AGK_1 zone catalogs. Similar work was done at the Cape Observatory (Cape of Good Hope) where proper motions for 34,000 stars were obtained by comparing the photographic positions with positions obtained from meridian catalogs of southern observations. All these proper motions have a comparatively low accuracy in view of the large errors of old meridian observations and, in the case of Yale catalogs, also due to the non-homogeneous system of the reference stars used in deriving the photographic positions. The accuracy of all these proper motions are within the limits of $\pm 0''.010$ and $\pm 0''.020$.

Among the catalogs there are the photographic catalogs of stars in the Kapteyn Selected Areas obtained at Pulkovo (18,000 stars) and at Radcliff (32,400 stars). At the McCormick Observatory proper motions were derived for stars from plates obtained for the determination of parallaxes.

Among the mass determinations of proper motions, one should point out the catalogs of proper motions derived from comparisons between the first epoch plates of the "Sky Map" and the astronegatives of the more recent epochs obtained with standard astrographs at the observatories of the Cape of Good Hope, Helsinki and Toulouse.

Relative proper motions for individual stars and groups of stars (for example, proper motions of star clusters, of the variable stars, etc.) are determined in great numbers at all observatories which have astrographs. Among the major catalogs, one should point out the Catalog of Proper Motions in the Pleiades published by Hertzsprung, and the Catalog of Stars with Large Proper Motions by Luyten. Separate determinations of proper motions are combined in the "Lexicon of Proper Motions" (EBL_2).

92. The Fundamental System of Faint Stars. The Work of Setting Up AGK_3R and the FK4 System

The compilation of fundamental catalogs in the last century and in the beginning of our century was basically necessary for the study of the motions of the bodies of the solar system and for performing geodetic work. Therefore bright stars were the principal contents of star lists of the fundamental catalogs, since they would be visible during the day time with meridian instruments and during the night with moderate field instruments.

The development of photographic methods necessitated the creation of fundamental systems based on fainter stars. The first step in extending the FK3 system to fainter stars was the compiling of the AGK_2A catalog. However, this catalog possesses a number of defects: it does not extend to the southern hemisphere, and the stars which are included have non-homogeneous characteristics. This circumstance is particularly inadmissible because systematic errors, which depend upon the stellar magnitude and spectral type, are inherent in photographic observations. The idea of creating a catalog of homogeneous, faint reference stars was suggested by Pulkovo astronomers in 1932 and in 1938, under the direction of M.S. Zverev, the realization of this plan was begun. It is known as the KSZ (Catalog of Faint Stars). The work of setting up a faint star catalog is divided into several tasks.

The first task is the creation of a fundamental catalog of faint stars—FKSZ. To this end there was compiled a list of 945 stars, evenly distributed over the whole sky, having visual stellar magnitudes between $7^m.5$ and $8^m.5$, and with small proper motions. The stars were chosen with spectral classes G and K so that they would have a numerically greater photographic magnitude. All observatories of the Soviet Union and some observatories of other socialist countries took part in making observations of these stars, in the declination zones down to -30°, for the first epoch. Thus more than twenty absolute and relative catalogs were obtained. From this group eight catalogs for right ascension and eight catalogs for declination were combined at the Pulkovo Observatory into a preliminary fundamental catalog of faint stars named PFKSZ. This composite catalog, set up in a system close to FK3, is of high accuracy due to the large numer of observations of each star (25 to 40). A number of observatories are completing the FKSZ observations by the absolute method. The inclusion of relative observations of the FKSZ within the FK3 system will permit future interconnection of the fundamental systems of bright and faint stars after the observations of the second epoch are complete and the FKSZ system established.

The second task entails meridian observations of the basic list of faint stars of KSZ in the FKSZ system. The stars of the catalog of faint stars were chosen so that in any area of 25 square degrees there would be 12 stars. The visual magnitudes of these stars are within the limits of $7^m.5$ to $9^m.1$; spectral classes range mostly from G to K. The total yearly proper motions of stars do not exceed $0''.04$. Thus the KSZ star list, containing 15,000 stars down to -30° in declination, is highly homogeneous (the mean visual magnitude is $8^m.5$, the color index is +1.1, which corresponds to spectral class G6, while the average total yearly proper motion is $0''.02$). It should be noted that several observatories in the southern hemisphere (e.g., the Cape Observatory) have expressed a desire to take part in the extension of the KSZ catalog to the southern sky.

All Soviet observatories and some foreign have joined in observations for the KSZ catalog. The preliminary reduction of the observations of the KSZ stars is carried out in the system of the PFKSZ catalog.

When the second epoch observations are completed, the final FKSZ system of positions and proper motions will be derived, which should then be improved with observations of minor planets and extragalactic nebulae relative to FKZ stars. This is the third task.

The program of observations of minor planets includes ten minor planets: Ceres, Pallas, Juno, Vesta, Hebe, Iris, Parthenona, Melpomene, Laetitia and Harmonia. The photographs are taken with the standard astrographs at Pulkovo and Tashkent as well as with the wide-angle astrograph at Moscow. Among foreign observatories taking part in the observations of minor planets are the observatories at Leiden, Sidney, Santiago, Cape of Good Hope and others. In view of the small field of the majority of astrographs, the positions of the minor planets are derived by the Schlesinger method, so far relative to the stars of the Yale catalogs and not relative to KSF stars (see Section 81). The precise theory of motion of the selected minor planets has been worked out at the Leningrad Institute for Theoretical Astronomy. One should note that a similar work of observing minor planets, using a slightly different program, is being conducted in the United States in accordance with a plan by Brouwer.

Extragalactic nebulae are observed in Kiev, Moscow, Pulkovo and Tashkent, using all available astrographs, as well as at Bucharest, Shanghai, Cordova and elsewhere. The program, arranged at Pulkovo and revised according to the number of photographs which have been taken, covers 157 areas, down to -5° in declination, which contain more than 450 galaxies down to $14^m.0$. The program for observing galaxies between -5° and -25° in declination was set up in Tashkent, and between -25° and -80° in Santiago, while for the region between -80° and -90° the photographs obtained at the Cape Observatory will be used at Pulkovo. For the purpose of possible tying in with the neighboring KSZ stars through the stellar background, the areas around KSZ stars are also photographed using small field astrographs. Taking of the photographs of extragalactic nebulae for the first epoch, with two positions of the instrument, can be assumed to be essentially finished. Extragalactic nebulae of the southern sky are photographed at the Cape Observatory and at the observatory of Santiago. We note that extragalactic nebulae are observed for astrometric purposes also at the Lick Observatory (see Section 82).

All work for the cataloging of faint stars should be finished before the end of the century.

Astrometric work associated with the new photographic reobservations of the zone catalogs and with the compilation of AGK_3 is clearly interconnected with the work for constructing the KSZ catalog. In virtue of the existence of the two catalogs AGK_2 and AGK_3 it will be possible to derive proper motions of stars and to calculate from them the stellar positions for epochs of old observations. Comparison of the calculated positions with the positions in AGK_1 will permit determination of the differences between the systematic errors in AGK_1 and in the new catalogs. In view of the inferior quality of AGK_1, these differences can be assumed to be errors of that catalog only. The combination of catalogs AGK_1, AGK_2 and AGK_3 will permit creation of a reference catalog of positions and proper motions for photographic observations, containing all stars down to $9^m.0$.

In order to facilitate the photographic observations of AGK_3, it was decided to establish a reference catalog of faint stars. Scott, at the Washington Observatory, compiled a list of about 14,000 stars. Some 30% of stars in this list were KSZ. Thus, by combining the KSZ and the Washington Observatory list into one catalog, a list of 21,505 stars was obtained; this reference catalog was designated as AGK_3R. Due to the large spread in the characteristics of the stars in the Washington list, the homogeneity of the AGK_3R stars is somewhat less than in KSZ.

In accordance with a resolution of the Ninth Assembly of the International Astronomical Union (Dublin, 1955), eleven observatories of the world, among them Pulkovo and Nikolayev, began observing the stars of the AGK_3R list between +90° and -5° in declination.

The FK4 fundamental catalog must serve as the fundamental system for the AGK_3R catalog. The list of stars of the FK4 catalog is somewhat larger than that of the FK3 catalog due to the addition of 1987 stars, primarily of magnitudes between $4^m.0$ and $6^m.0$. At the present time, corrections with respect to accidental and systematic errors for all stars of FK3 have been already published, thus bringing the coordinates of these stars within the coordinate system of FK4. The work of compiling the FK4 fundamental catalog was completed in 1963.

In concluding this text, devoted to a short exposition of the problems encountered in determining the coordinates of celestial bodies, we should note that the primary task of the astrometrist—the creation of a highly accurate system of celestial coordinates—can be solved only on the basis of friendly collaboration of observatories throughout the world. It will also require many years of painstaking work by amateur observers, each making his small contribution to the vast collection of observations. Not all of these contributors will see the ultimate creation of this new, most precise system of coordinates. However, the fruits of many decades of work by astrometrists will serve to further the progress of humanity toward a knowledge of the secrets of the universe.

Bibliography

CAMPBELL, W. W. The Elements of Practical Astronomy. Macmillan, New York, 1899.

CHAUVENET, WILLIAM. A Manual of Spherical and Practical Astronomy. Vol. II, Chapters V, VI, and VII. (5th Edition, reprinted by Dover, New York.)

DIECKVOSS, W. Systematic Errors in the AGK2 and Final Reductions in the AGK3 Program. Astronomical Journal. 67:686, 1962.

FRICKE, W. and A. KOPFF. Fourth Fundamental Catalogue (FK4). Veröffentlichungen des astronomischen Rechen-Instituts, Heidelberg. No. 10. The introduction (in English) describes all the catalogs which were combined to form this catalog.

NEWCOMB, SIMON. A Compendium of Spherical Astronomy. Chapters XII and XIII. Macmillan, New York, 1906. (Reprinted by Dover.)

NEWCOMB, SIMON. The Elements of the Four Inner Planets and the Fundamental Constants of Astronomy. Supplement to the American Ephemeris and Nautical Almanac for 1897. Washington, 1895.

REPSOLD, I. A. Zur Geschichte der astronomischen Messwerkzeuge. Vol. I and II. Leipzig, 1908.

SCHLESINGER, FRANK. Photographic Determinations of Stellar Parallax. In: Probleme der Astronomie. Festschrift für Hugo v. Seeliger (pp. 422-439). Springer, Berlin, 1924.

SCOTT, FRANCIS P. Status of the International Reference-Star Program. Astronomical Journal. 67:690, 1962.

SMART, W. M. Textbook of Spherical Astronomy. Chapters IV, VIII, IX, X, XI and XIII. Cambridge University Press, London and New York (latest ed.).

TURNER, H. H. Preliminary Note on the Reduction of Measures of Photographic Plates. M.N.R.A.S. 54:11-25, 1893.

ZVEREV, M. S. Fundamental Astrometry. Translated for the U. S. Naval Observatory. U. S. Dept. of Commerce, Office of Technical Services, Washington, 1963 (OTS no. 63-11, 569).

New Instruments and Methods in Meridian Astrometry. Reports made at the 10th Assembly of the International Astronomical Union, Moscow, 1958. All papers are printed in two languages (Russian, English, French, German). USSR Academy of Sciences Press, Moscow, 1959.

Subject Index

Index of Names